Springer Optimization and Its Applications

Volume 175

Aims and Scope

Optimization has continued to expand in all directions at an astonishing rate. New algorithmic and theoretical techniques are continually developing and the diffusion into other disciplines is proceeding at a rapid pace, with a spot light on machine learning, artificial intelligence, and quantum computing. Our knowledge of all aspects of the field has grown even more profound. At the same time, one of the most striking trends in optimization is the constantly increasing emphasis on the interdisciplinary nature of the field. Optimization has been a basic tool in areas not limited to applied mathematics, engineering, medicine, economics, computer science, operations research, and other sciences.

The series **Springer Optimization and Its Applications (SOIA)** aims to publish state-of-the-art expository works (monographs, contributed volumes, textbooks, handbooks) that focus on theory, methods, and applications of optimization. Topics covered include, but are not limited to, nonlinear optimization, combinatorial optimization, continuous optimization, stochastic optimization, Bayesian optimization, optimal control, discrete optimization, multi-objective optimization, and more. New to the series portfolio include Works at the intersection of optimization and machine learning, artificial intelligence, and quantum computing.

Volumes from this series are indexed by Web of Science, zbMATH, Mathematical Reviews, and SCOPUS.

More information about this series at http://www.springer.com/series/7393

J. MacGregor Smith

Combinatorial, Linear, Integer and Nonlinear Optimization Apps

COLINA Grande

 Springer

J. MacGregor Smith
Department of Mechanical
and Industrial Engineering
University of Massachusetts
Amherst, MA, USA

ISSN 1931-6828   ISSN 1931-6836   (electronic)
Springer Optimization and Its Applications
ISBN 978-3-030-75803-5   ISBN 978-3-030-75801-1   (eBook)
https://doi.org/10.1007/978-3-030-75801-1

Mathematics Subject Classification: 65K05, 90C05, 90C10, 90C11, 90C27, 90C29, 90C30, 90C59, 90C90, 68W01, 68W05, 68-04

This Springer imprint is published by the registered company Springer Nature Switzerland AG
The registered company address is: Gewerbestrasse 11, 6330 Cham, Switzerland

Dedicated to my loving wife Marty and our daughter Kate for always providing me support and encouragement during the development of this volume. Also, to the students of MIE 379, 380, and 724 and other courses who shared my enthusiasm for developing the Apps along with their excruciating efforts to understand the problems and code them.

Preface

We shape our tools, thereafter our tools shape us.

—MARSHALL MACLUHAN

Personal mobile computing has profoundly altered the workplace during the past decade. Apps on phones and tablets are commonplace and are becoming more important for the workplace and everyday life. We need optimization Apps which are not just descriptive of a situation but also prescriptive, so that we can make better decisions. These are more challenging than simple descriptive Apps and more rewarding in the long run.

Our goal is for users to be creative consumers of the new technology. How can we benefit and realize this process and how can we teach our students at the university level how they can contribute to making things and making them better, no matter what engineering, science, or business discipline they originate from?

On a personal note, the origins of my interest in Apps came from reading the Business section of the Sunday New York Times in 2010 about a programming class at Stanford where students were learning how to program a phone App on Apple iPhone which they could market. I asked why can't we develop a similar program at the University of Massachusetts? I found out about App Inventor Classic version #1 which was just being developed at Google.

In the Fall of 2011, I started teaching about App Inventor in my MIE 379: *Introduction to Operations Research* class and had the students carry out an App Inventor optimization term project. I felt that creating an optimization App would be challenging and give our students a unique slant on the App market.

It was crazy ambitious since the App Inventor programming language was in its initial developing stages. The thing that grabbed my attention was the ability of App Inventor to design the user interface for the phone with all the buttons, labels, text, and diagrams. The materials in this book reflect the lessons learned from my classes. In addition, two spinoff App products derived from App Inventor have been developed and we will illustrate these new developments at the end of this volume which are Thunkable and Kodular. These additional software developments were intended to make AI2 and its extensions more powerful and applicable to other than educational objectives. Thus, the future of App development based upon App Inventor is on an upward trajectory.

0.0.1 Why this Volume

Why are Operations Research and Computer Science relevant and important? Operations Research (OR) is relevant and important because almost all optimization problems require some theoretical and applied mathematics understanding as their foundation. The formulation of optimization problems is founded through OR concepts and techniques: *Combinatorial Optimization, Linear Programming, Integer, and Nonlinear Programming (COLIN)*. Computer Science (CS) is relevant and important because algorithms and *Apps/algorithms (A)* underlie the solution of all optimization problems. App Inventor and AMPL are the major algorithmic vehicles studied in this course. Thus, in summary we

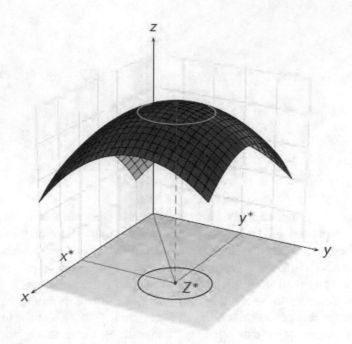

Fig. 0.1 COLINA grande optimization (after PSTricks website tug.org)

have COLINA which means "small hilltop" in Spanish and it is Grande! Figure 0.1 illustrates a small hilltop optimization problem in the decision variables x and y.

For whom is this book written? University-level students and professors, mainly at the upper division level or higher with some background in calculus, linear algebra, and related mathematical skills. These skills are viewed as sufficient but not necessary for developing the Apps for Optimization, *i.e.* OptApps or AppOpts.

0.0.2 Plan of the Volume

How is this book organized? Basically, each chapter provides the theoretical background and the general methodology for the concepts of optimization according to COLINA, then there are example applications of the Apps with App Inventor and AMPL software. The COLINA categories are representative of the major areas of interest in optimization which I have come across in my teaching experience. Links to the University of Wisconsin NEOS server which is designed to solve complex optimization problems are used in some of the Apps where more computing firepower is required. All Apps developed in the book are available for download.

Amherst, MA   *J. MacGregor Smith*
2020   *University of Massachusetts*
                                                                        Amherst, Massachusetts

Acknowledgements

I am indebted to Italo de Souzsa (class of 2016) for his creative insights on how to link App Inventor 2 to the NEOS server in Wisconsin. This has been instrumental in extending the capabilities of AI2 to larger and more complex optimization problems. I am also deeply indebted to Ben Huebner and Elizabeth Wong of the University of Wisconsin Department of Computer Science who designed a way to update the communications with the NEOS servers when the email connection was removed.

I would like to also thank Peter Kubat for being a good friend, critic, and supporter over the years and for his encouragement about the Apps for Optimization. I would also like to thank Hal Ableson and his colleagues at MIT for developing App Inventor and Robert Fourer for his help with AMPL.

I would like to thank Dr. Mei-Yau Shih of the University of Massachusetts Amherst Center for Teaching (CLT) for her enthusiastic encouragement and financial support in the development of the Apps in my courses.

I would like to thank the following students who contributed apps to this volume (in order of appearance of their Apps): Connor Tremarche, Amanda Skriloff, D. Sokol, Jared Marvel, Ian Taylor, Avery Stroman, Ryan Barnes, Alex Barth, Anthony Broding, Sydney Hauver, Alexander Niemeyer, Brendan Frakfort, Thomas Johnson, Andrew Metz, Bobby Jaycox, Tim Klocker, Eric Wright, Tom Rogers, Allan Tang, Erin O'Neil, Bekah Perlin, Lily Thomas, Thimi Prifti, Rene Arnaud, Crystal Lee, Andrew Giampa, Joe Woodman, Johnny Zhu, Milagros Malo, Rebecca Castonguay, Sean Fitzgerald, William Andrews, Prashant Meckoni, Xi Jiang, Luyi Wang, and finally, Rahda Dutta. Rahda was also very helpful in providing background materials and instructions. Nicole Lynch also helped out with instructions on using AMPL.

I would also like to thank the anonymous referees who read over the book drafts and provided valuable comments on the manuscript. I would also like to thank Donna Chernyk of Springer for her continued support and encouragement in developing the book.

Finally, I would like to thank all the undergraduate and graduate students of the University of Massachusetts at Amherst who have participated in my courses on Linear, Network, and Nonlinear Programming over the years.

Contents

Acronyms

AHP	Analytical Hierarchy Process
AMPL	A Mathematical Programming Language
BFS	Basic Feasible Solutions
CO	Combinatorial Optimization
COLINA	Combinatorial, Linear, Integer, and Nonlinear Optimization Apps
DTM	Deterministic Turing Machine
EOQ	Economic Order Quantity
ERT	Equalization of Runout Times
ILOG	Intelligence and Logic, a software company
ILP	Integer Linear Program
IP	Integer Program
LP	Linear Program
MIP	Mixed Integer Programs
MST	Minimum Spanning Tree
NATO	North Atlantic Treaty Organization
NDTM	Non-deterministic Turing Machine
NEOS	Network-Enabled Optimization System
NLP	Nonlinear Programming
QAP	Quadratic Assignment Problems
SMT	Steiner Minimal Trees
SPT	Shortest Path Tree
TM	Turing Machine, after Alan Turing, computer scientist
TND	Topological Network Design
TP	Tame Problems
TSP	Traveling Salesman Problem
VPL	Visual Programming Language
WP	Wicked Problems

1

App Problem Formulation

Overview In this chapter, we give the reader an overview of problem formulation and its relationship to App development and how Operations Research and Computer Science strategies are helpful in constructing methodologies for their solution. See Figure below.

- Optimization App Problems (OptApps)
- Wicked Design Problems (WDPs)
- Personal Mobile Design Problems (PMDPs)

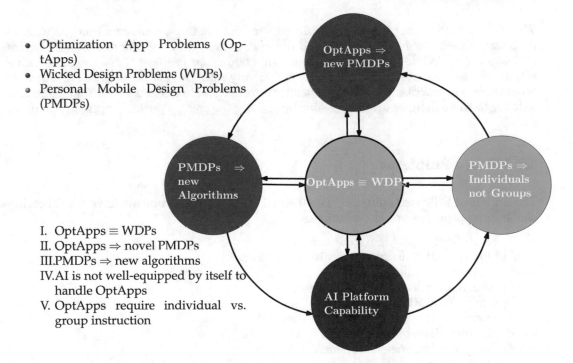

I. OptApps ≡ WDPs
II. OptApps ⇒ novel PMDPs
III. PMDPs ⇒ new algorithms
IV. AI is not well-equipped by itself to handle OptApps
V. OptApps require individual vs. group instruction

Keywords: Problem Formulation, Tame Problems, Wicked Problems,

© Springer Nature Switzerland AG 2021
J. MacGregor Smith, *Combinatorial, Linear, Integer and Nonlinear Optimization Apps*,
Springer Optimization and Its Applications 175,
https://doi.org/10.1007/978-3-030-75801-1_1

1.1 Prologue

> The mere formulation of a problem is far more essential than its solution, which may be merely a matter of mathematical or experimental skill. To raise new questions, new possibilities, to regard old problems from a new angle requires creative imagination and marks real advances in science.
>
> —ALBERT EINSTEIN

> A powerful programming language is just a means for instructing a computer to perform tasks. The language also serves as a framework within which we organize our ideas about processes.
>
> —HAL ABELSON

> Successful problem solving requires finding the right solution to the right problem. We fail more often because we solve the wrong problem than because we get the wrong solution to the right problem.
>
> —R.L. ACKOFF

1.2 Introduction

We need to realize that most real-world design and planning problems that we wish to optimize are extremely challenging and difficult, in fact, they are often viewed as *Wicked Problems (WPs)*. While in school, we are taught about *Tame Problems (TPs)*, the ones with a single solution, but in reality there are usually many solutions and thus, WPs are normally what we face in practice. However, we need to approach these problems with algorithms and a desire to solve them as best as possible by seeking the crest of the local optimal solutions.

1.3 Wicked Problems

The notion of WPs was coined by Horst Rittel [11]. Wicked problems have the following challenging properties or characteristics:

- $\mathscr{P}_1$ No definitive problem formulation.
- $\mathscr{P}_2$ No exhaustive list of permissible operations.
- $\mathscr{P}_3$ No stopping rule.
- $\mathscr{P}_4$ No **single** criterion for correctness.
- $\mathscr{P}_5$ Many **alternative solutions**.
- $\mathscr{P}_6$ Every WP is symptomatic of another WP.
- $\mathscr{P}_7$ No immediate or ultimate test of a solution.
- $\mathscr{P}_8$ Every WP is a one-shot operation.
- $\mathscr{P}_9$ Every WP is essentially unique.
- $\mathscr{P}_{10}$ We are morally responsible for a solution.

The set of properties of a WP indicate its richness and difficulty. $\mathscr{P}_1$: *No definitive problem formulation* means that you cannot simply write down the problem description on a sheet of paper, give it to somebody, and they can go off in the corner and solve it. It also implies that a proper problem formulation indicates the actual solution, however, there is no single way to define the problem since it depends upon your worldview. Your worldview not only helps you understand the problem, but it also colors your perspective. C. West Churchman, the famous systems scientist, was instrumental in gathering people together so that many worldviews would be present in resolving system planning problems, not just one.

In this book, we have gathered together a number of different problems and applications to illustrate the variety of approaches in problem-solving.

$\mathcal{P}_2$: *No exhaustive list of permissible operations* means that there is no exhaustive set of steps to guarantee a solution. We could read all the books on a subject, ask all the important people affected by the problem, sample and observe the situation, and so on. There are many alternative methods and ways to approach a problem concomitant with the different worldviews. A person's worldview is crucial to the approach for solving the problem.

$\mathcal{P}_3$ *No stopping rule* implies that you cannot stop the solution process because you can always do better. This is most frustrating for decision makers since the word *decidere* comes from the Latin phrase "to cut off". Normally, people give up because they have exhausted all their resources.

$\mathcal{P}_4$ *No* **single** *criterion for correctness* indicates that the problems are multi-objective in nature involving often intricate complicated trade-offs. There can be many alternative solutions since the problem is multi-objective and has a complex noninferior set of solutions[1]. The noninferior set can have an infinite number of solution alternatives, not just a single one.

$\mathcal{P}_5$ *Many* **alternative solutions** are concomitant with the multi-objective nature of the problem; there are many alternative solutions, not one best solution.

$\mathcal{P}_6$ *Every WP is symptomatic of another WP* means that the WPs are interdependent and nested together and not separable. Normally, we like to decompose a problem into separate pieces, but WPs are anathema to this strategy.

$\mathcal{P}_7$ *No immediate or ultimate test of a solution* implies that there is no immediate or ultimate way to guarantee a solution's correctness. There is no simple simulation model to test the robustness of the solution. The fact that there are many alternative solutions and no single criterion for success means that we have no guarantees.

$\mathcal{P}_8$ *Every WP is a one-shot operation* usually means that the resources needed to solve a WP imply that you only have one chance to solve it without major consequences for adjusting your solution. Normally, the expenses of tearing down and starting over far outweigh that strategy.

$\mathcal{P}_9$ *Every WP is essentially unique* generally means that each WP is unique in time and space. We cannot learn by solving WPs over and over; we must start anew again. This is the most frustrating property since we surmise from past experience that our learning about problem-solving provides a straightforward framework and underpinning for solving new problems. But this is just magical thinking; there are no experts.

$\mathcal{P}_{10}$ *We are morally responsible for a solution* implies that we are ethically responsible for our actions and we have to be morally responsible in our decision-making. In engineering, architecture, law, and most professional disciplines, we have an obligation to society to provide the best advice possible.

While real-world problems are WPs, we still have to work through them. This is why we need computing help and algorithms in order to deal with WPs.

As an example of a WP, let's examine the process of selecting a golf club to hit a golf shot. This is seemingly a trivial problem to the non-golfer ("it's only a game"), but in reality, for the average golfer this problem is most complex. Even for the pros, this is challenging, and it is why they are so inconsistent. Why is this the case? It really stems from the fact that each time the ball lands, it ends up in a different lie situation, so it makes the problem unique. The ball can be located in a wet sand trap, on the downside or upside of a hill; there is moisture or dirt on the ball; the ball lies on bare ground; there is a twenty-mile and hour wind blowing in the golfer's face; and so on. Representing the actual situation requires critical information, and there is a great deal of *uncertainty and variability* as to how to assess the situation.

---

[1] Noninferior for a multi-objective problem means that there exists no other feasible solution that will yield an improvement in one objective without causing a degradation in at least one other objective. [4]

- What is the true distance to the pin when the green is situated 10 feet above your position on the course?
- In match play, should you play safe or take a risky shot, since your partner's situation also becomes a factor to consider?
- In the wet rough, should you use an iron or a fairway wood because the fairway wood will be likely be impeded by the moisture?
- Should one play over or under the trees?
- Should you bump and run or fly it to the pin from the fairway?

There is only one chance for success. How can we design a phone App to assist the average golfer? How can Computer Science help us with algorithms for solving WPs?

Computer Science has developed a set of problem categories that are useful in designing algorithms for these problem solutions. It is a very useful characterization because if one knows the category of the problem, then one can assess the difficulty of the problem solution and an appropriate algorithm.

The problem categories are based on a fundamental abstract model of computation called the Turing Machine (TM) model of computation, or Deterministic Turing Machine (DTM) for short. A DTM is a model representing most practical, everyday computers. Another useful abstract comparison model is called a Non-deterministic Turing Machine (NDTM) for short. A NDTM does not really exist; it is fictitious, but it is helpful in the problem classifications. Briefly, here are the complexity classes:

- $\mathscr{P}$, The class of problems which are solvable by a DTM in polynomial time.
- $\mathscr{NP}$, The class of problems which are solvable in polynomial time on a NDTM.
- $\mathscr{NP}$-Hard, The class of problems at least as hard as the hardest problems in $\mathscr{NP}$.
- $\mathscr{NP}$-Complete, The class of problem in $\mathscr{NP}$ and the intersection of $\mathscr{NP}$-Hard which are transformable in polynomial time to the *3-Satisfiability* problem.
- *Undecidable* are those problems where it is impossible to construct an algorithm (TM) that always leads to a correct yes-or-no answer. In effect, the Undecidable problems are almost equivalent to the Wicked Problems described earlier, but WPs remain worse than the Undecidable ones to solve.

Figure 1.1 illustrates the relationships between the problems of Computer Science and those problems we call Wicked Problems. Most of the problems we examine in this book will be from the class $\mathscr{P}$ simply because of the phone or tablet's limited computational power. Sometimes, an $\mathscr{NP}$-Hard problem can be solved optimally if the size of the problem is small. That is why the link to the NEOS server is provided, but still, Wicked Problems are truly difficult. The key issue facing us is: *Can we solve WPs with systematic procedures?*

1.4 Systematic Procedures and Algorithms

In this course and book, we will examine COLINA-type problems with basically nine systematic procedures and algorithms for their solution. The range of algorithms is appropriate depending on the particular type of problem. Many problems will be solved by combining some of the strategies. The idea is to widen our approaches for solving problems rather than to rely only on a singular technique such as Linear Programming (LP). We give a brief definition of the different techniques.

- **Sorting and Searching:** Sorting and search methods are very basic and should be included first as they are classical procedures that are part and parcel of an algorithm library and can be very useful in different applications.

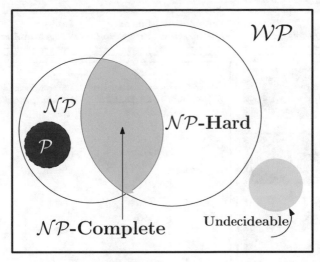

Fig. 1.1 Relationship between Computing and Wicked Problems

- **Greedy Solutions**: Solving a problem by carefully selecting components of the problem in a greedy fashion and putting them together for a solution of the entire ensemble is very sensible, but is normally a heuristic approach.
- **Divide-and-Conquer** (*a.k.a* **Branch and Bound**): Partitioning a problem into different sets and recursively constructing a solution for integrating the elements within the partitions.
- **Backtracking or Enumeration**: Essentially, a process for identifying and enumerating all the possible combinatorial solutions of a problem. Normally, all the different solutions are represented as a tree and the solutions are developed by moving down the branches of the tree and backtracking to the top of the tree and searching for a new path again until all solution paths have been enumerated. While pedestrian in nature, it is often the only method of solution.
- **Calculus**: Use of the calculus to identify the derivatives of a continuous problem, setting the equations to zero, and finding their solution. Second derivatives are useful for determining the convexity of functions. Solving the resulting equations which are often nonlinear themselves can remain challenging.
- **Linear Equations**: Developing a set of linear equations for a problem and solving them by Gaussian methods, Gauss-Jordan, or Chelosky decomposition.
- **Linear Programming (LP)**: Essentially setting up a linear objective function and constraints for a problem and using standard methods such as the simplex or dual simplex method for their solution. This is one of the basic methods taught in the MIE 379 course.
- **Dynamic Programming**: Use of Bellman's equations and the concepts of state space and stages in a recursive approach to enumerate solutions to a problem. Certain problem types which have a recursive nature to them lend themselves to this technique. For example, shortest paths are a prime candidate of his approach.
- **Nonlinear Equations and Numerical Programming (NLP)**: Use of calculus to generate derivatives and find the optimal solutions or numerical techniques and Lagrangian techniques to solve a set of nonlinear equations.
- **Computational Geometry**: Use of convex hulls, geometric searching, Voronoi diagrams, or Delaunay triangulations to solve a geometric optimization problem. Location optimization problems are very amenable to this approach.

Within the book and in the various application discussions, we shall illustrate the strategies and algorithms within the App rather than as a separate entity. It is felt that the strategies are best seen within the application domains. Table 1.1 illustrates this concept.

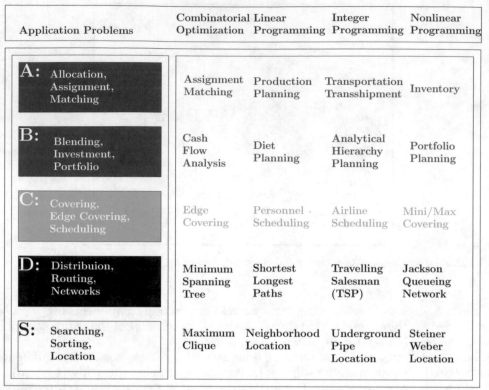

Application Problems	Combinatorial Optimization	Linear Programming	Integer Programming	Nonlinear Programming
A: Allocation, Assignment, Matching	Assignment Matching	Production Planning	Transportation Transshipment	Inventory
B: Blending, Investment, Portfolio	Cash Flow Analysis	Diet Planning	Analytical Hierarchy Planning	Portfolio Planning
C: Covering, Edge Covering, Scheduling	Edge Covering	Personnel Scheduling	Airline Scheduling	Mini/Max Covering
D: Distribuion, Routing, Networks	Minimum Spanning Tree	Shortest Longest Paths	Travelling Salesman (TSP)	Jackson Queueing Network
S: Searching, Sorting, Location	Maximum Clique	Neighborhood Location	Underground Pipe Location	Steiner Weber Location

Table 1.1 App Strategies and Problems

There are many caveats to mention here since App Inventor itself is not a traditional sophisticated textual programming language. For optimization, it has no libraries, so things like sophisticated data structures (*e.g. matrices*) or sorting routines do not exist *a priori*; one has to build them from scratch. This is problematic but not insurmountable, however, one has to be realistic in that working through the available resources is the bottom line. AI2 has no output software graphics library itself and graphics must be imported, but the programming blocks are very visual and nicely designed which is a real plus. Also, as has been mentioned, when more computing power is needed there is always the NEOS server which can be used to solve complex optimization problems as we will show. However, it is best if the App can be self-contained where it will rely solely on the computational power of the phone. So the problem formulation of the App is really crucial.

1.4.1 Plan of the Book

The plan of the book is to introduce each App in a concise framework as best as possible with the following organizational structure:

- Introduction
- Problem
- Mathematical Model
- Algorithm
- App Demonstration
- Evaluation (Benefits and Costs)

1.4.2 App Design Strategy

App Inventor basically has three main steps or event functions in the App development:

- Designer Screen Editor
- Blocks Programming Editor
- Testing App on Phone or Emulator

Fig. 1.2 Designer Screen Layout

First, one must design the layout of the screen with all the buttons, divisions of space, and labels. This is often the fun part and the selection of the items to be included on the phone is pretty rich. Figure 1.2 illustrates the screen design dialog for generating our first App in Chapter 2. App Inventor has many possible inputs and arrangements for designing the screen which makes it quite flexible and powerful.

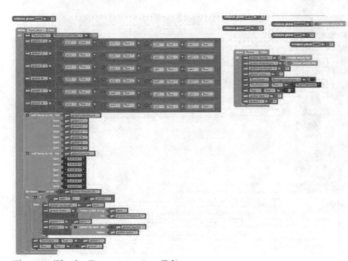

Fig. 1.3 Blocks Programming Editor

Figure 1.3 illustrates part of the blocks programming for our first App. The second part deals with the programming of the visual blocks. This is usually the hard part, which includes the algorithmic strategy and also any debugging. AI2 does a very good job of organizing the blocks according to their functions, with color-coded shapes and easy maneuverability on the screen, as opposed to traditional text-based programming languages.

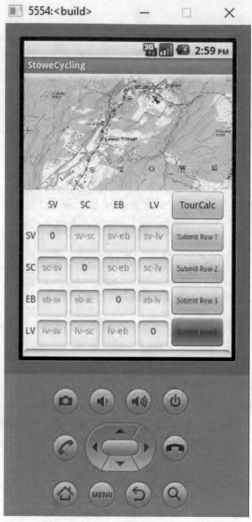

Fig. 1.4 Testing App on Phone or Emulator

The third part is the demonstration of the App with the phone emulator or actual smartphone or tablet. This also can be fun if done properly. It is probably best if one has an Android phone for AI2 because then the interface between the blocks programming and the user is seamless. AI2 is considering developing a version that works for IOS-type phones, but it is still in the works. Figure 1.4 illustrates the phone emulator for our first App. This phone emulator available in AI2 is subject to connection problems, so it is probably best to use the smartphone itself to examine the results of the App although sometimes it is more convenient to use the emulator. Since the Apps are designed to be optimization tools within the context of our COLINA viewpoint, the structure of the Apps is somewhat unique because we are basically developing mathematical models for solving optimization problems.

1.4.3 COLINA Methodology

Within App Inventor 2, the strategy we normally follow in programming the blocks is roughly according to the following flowchart representation of the general approach for creating an AI2 App for optimization (Fig. 1.5).

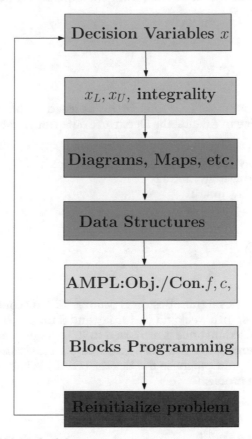

Fig. 1.5 Flowchart of COLINA Methodology

- Identify the Decision Variables and Parameters. If possible, talk to the client(s) about their needs and objectives.
- Initialize their values, both lower and upper bounds if necessary.
- Utilize diagrams, flowcharts, or graphs to illustrate the optimization problem and its processes.
- Collect all necessary data and set up any necessary list data structures, matrices, or tables.
- Identify the Objective Function and any Constraints (Equalities and Inequalities). Eliminate variables from the constraints whenever possible, and use back substitution to simplify the expressions. Run a test example of your model with AMPL to make sure the elements of the problem will execute. AMPL is suggested since it will work with the NEOS server.
- Develop procedures to process the variables according to the desired strategy for solving the problem. These are repeatable code segments called iteratively during the processing.
- Set up any restarting procedures to reset the variable values and iteratively test and re-run the App.

Let's demonstrate the generation of one of the Apps according to our COLINA framework with AI2.

1.5 Airline Tickets

As a simple example of an App, let's consider the problem of determining the best way to travel from Boston to Atlanta or to any other pair of cities where airline discounts apply over a 4-week period. This is a type of transportation-assignment problem. This problem was discussed in Taha's textbook, but one can generalize it and make it more useful with an App [14].

1.5.1 Introduction

A business executive must make four round-trips as listed in the following table between Boston and a branch office in Atlanta: the Departure date from Boston, and the Return Date from Atlanta

a. Monday, June 3; Friday, June 7
b. Monday, June 10; Wednesday, June 12
c. Monday, June 17; Friday, June 21
d. Tuesday, June 25; Friday, June 28

1.5.2 Problem

The price of a round-trip ticket from Boston is $400. A 25% discount is granted if the date of arrival and departure span a weekend (Saturday and Sunday). If the stay in Atlanta lasts more than 21 days, the discount price is increased to 30%. A one-way ticket between Boston and Atlanta (in either direction) costs $250. Formulate an assignment/matching problem for how the executive should purchase the tickets. (**Hint:**) What are the sets (pairings) of decision variables in the problem?

1.5.3 Mathematical Model

The modeling trick in this problem is to look at the pairing of the origin and destination of the flights over the given four dates of the one-month period incorporating the discounts as a function of time in weeks. These are the decision variables of our problem. This data can be set up as a matrix data structure.

1.5.4 Algorithm

We solve the problem by enumeration as there is no direct formulation of the problem as an obvious assignment/matching problem.

1.5.5 App Demonstration

Once we input the parameters on the costs and discounts, the algorithm enters the overall cost for the combinations of arrivals/departures as a function of time in the cost matrix. As you can see in the matrix screen solution on the right, the *21-day discount* results in the least expensive way to travel. So our formulation of the problem leads to its solution (Figs. 1.6 and 1.7) for the blocks.

Fig. 1.6 Input and Parameter Setup with Solution Matrix

1.5.6 Evaluation (Benefits and Costs)

The App is fairly straightforward and once the parameters and output numbers are illustrated in the matrix, one can read off the best solution. It is not only a very simple App to use but also effective.

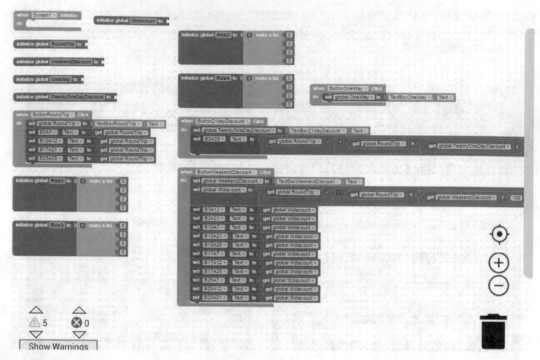

Fig. 1.7 Airline Blocks

1.6 A Brief History of Visual Programming Languages

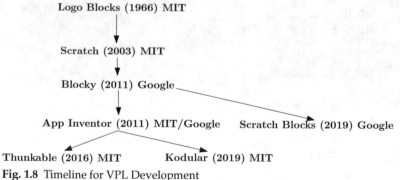

Fig. 1.8 Timeline for VPL Development

Visual Programming is a relatively recent phenomenon largely aimed at encouraging new users to learn coding without the traditional textual basis found in most computer programming languages. Definition: *A Visual Programming Language (VPL) is any programming language that lets users create programs by manipulating program elements graphically rather than specifying the elements "textually" such as in Java, Python, and C.* Visual expressions, spatial arrangements of coding statements of combined text, and graphic symbols are part of a VPL. A flowchart language with "boxes and arrows" is a good VPL example. Visual blocks coding is especially useful for children who cannot read or type very well, but it is not limited to young adults. Many programming languages have drag-and-drop blocks or modules. A block's shape and color give an indication of its purpose. The way in which the blocks are assembled shows the flow of the program. Categorization by color increases one understanding of what the various program elements are doing (Fig. 1.8).

The first VPL was LOGO Blocks (1966) founded at MIT. The next most important VPL SCRATCH (2003) also developed at MIT at the Lifelong Kindergarten Laboratory, then in 2011, Google created BLOCKY which generates code in other languages from the visual blocks. SCRATCH BLOCKS is a current derivative development from SCRATCH and BLOCKY (2019). After BLOCKY, we have App Inventor Classic which was a joint project between MIT and Google in 2011. This language and its extensions will be the fundamental foremost one used in this text. Finally, there are the derivative languages Thunkable and Kodular which emanate from programmers who worked at MIT with App Inventor 2.

As we will demonstrate, AI2 has advantages and disadvantages. The advantages stem from its user visual interface and blocks programming features. AI2, however, was only developed for Android phones, so it leaves out Apple (IOS) phones. AI2 may change this omission in the near future. This omission has been remedied with Thunkable which allows IOS phone systems. Kodular is developing a version for IOS phones.

Both Thunkable and Kodular allow for the inclusion of spreadsheets through outside vendors which is useful for large data sets; AI2 does not have this outside spreadsheet feature. AI2 is free and open-source whereas Thunkable and Kodular are not. Kodular has more features in its interface and is more complex in its options than Thunkable. Table 1.2 summarizes the advantages and disadvantages of the software tools.

One may wonder why I did not include Apps from Thunkable and Kodular. I thought about that, but then realized that the basics of AI2 are the foundation of these latter tools which are still under development. So grounding the development in AI2 is essential to understanding Thunkable and Kodular.

Code	Oper. Systems	Open Source	Features	Data Inputs	Cost
AI2	Android	Open Source	Basic Features	Fusion Tables	Free
Thunkable	Android + IOS	No	Basic$^+$ Features	Spreadsheets	Free + Pro$
Kodular	Android + IOS	No	Advanced Features	Spreadsheets	Free + $

Table 1.2 Advantages and Disadvantages of VPL for COLINA

1.6.1 Kodular Demonstration

Figure 1.9 illustrates the input screen and the solution for the required parameters in Kodular. Notice that the Designer Screen has a similar layout and organization to the AI2 input screen of Figure 1.5. The colors are slightly more brilliant in Kodular and more interesting than AI2.

The visual programming blocks of Kodular look very similar to those of AI2 In Figure 1.6, Kodular as well as Thunkable are grounded in AI2's design structure and in addition offer extensions to the basic commands of AI2. So once you understand AI2's logic and organization, you are well-equipped to tackle these other two App creators.

Figure 1.10 illustrates Kodular's Blocks.

1.7 AI2 Introduction

To set up

1. Create a Google gmail account, if you do not already have an account, and start working with AI2, you will need to do the following. Google will assist in your log on to App Inventor (AI) if you have a gmail account.
2. You may use either a PC or a Mac to run App Inventor.
3. Chapter 2 in Kamarani and Roy: *App Inventor 2 Essentials* shows in detail how to set up AI2.

Fig. 1.9 Kodular Input Setup with Background Commands

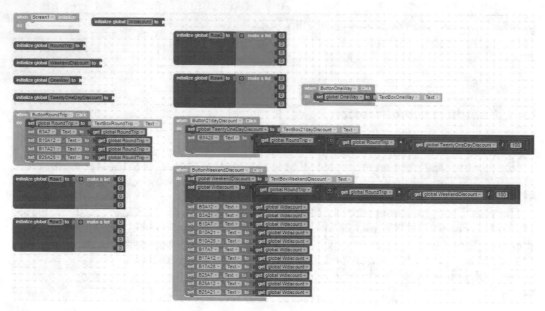

Fig. 1.10 Kodular Airline Blocks

4. Go the the MIT App Inventor Program website and read the instructions on how to start App Inventor. http://www.appinventor.mit.edu

   o You need to download a file that accesses App Inventor (AI).
   o You should make sure you also have the **latest** java code (version 7+) to get classical AI to work smoothly.
   o **David Wolber's book** www.appinventor.org/book2 is very useful.

5. There is a tutorial site for the book which helps create the Apps from the book. http://www.appinventor.org/

6. There are many other tutorial sites with more examples, called Puravida Apps: http://puravidaapps.com/learn.php There are many tutorials on this site.
7. There is a dynamic video for some very well-done and interesting examples which are highly recommended. Krishnendu Roy's website (*App Inventor 2 + Classic examples*): http://coweb.cc.gatech.edu/ice-gt/1646

1.8 Exercises

The following exercises complement this chapter.

1. **Wicked Problem Example:**
   Take the key characteristics of a Wicked Problem (WP) as described in the text and generate what you feel is an example of a WP delineating your example against all the characteristics.

   The next exercise is related to the Airline Tickets example just presented, but only requires a graphical solution, not an App itself.
2. **Airline Scheduling Problem:**
   Imagine you have a 5-week business commitment between Boston, MA (BOS), and San California (SFO). You fly out of BOS on Mondays and return on Wednesdays. A regular round-trip ticket costs $500, but a 15% discount is granted if the dates of the ticket in either direction cost 75% of the regular price. How should you buy the three possible alternatives are presented in a), b), and c):

   a. Buy 5 regular BOS-SFO-BOS tickets for departure on Monday and Wednesday of the same week.
   b. Buy one BOS-SFO, 4 SFO-BOS-SFO that span weekends, and one BOS-SFO ticket.
   c. Buy one BOS-SFO-BOS to cover Monday of the first week and Wednesday of the last week and 4 SFO-BOS-SFO to cover the remaining legs.

   - All tickets in this alternative span at least one weekend.
   - The restriction on these options is that you should be able to leave BOS on Monday and return on Wednesday of the same week.

   Develop a graphical solution to the problem. **A Gantt chart** is a good approach (Fig. 1.11). The Gantt chart shows the scheduling of three jobs on a machine over time and their precedence relationships.

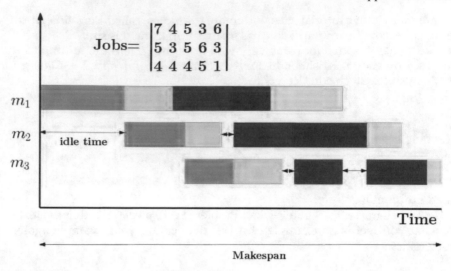

Fig. 1.11 Gantt Chart

3. **Golf Club Selection Problem:**

Develop an App to help the average golfer choose his/her clubs. Develop a data distance chart with a range of distances for the woods and irons from the driver through the pitching wedge which the user would enter. Also, develop a set of characteristics of the lie of the ball that might increase or decrease the percentage yardage that the golfer would use to select the right club. For example, here are some estimates of the range for our clubs: [2]

Driver :   200–240 yds (183–219m)
3-Wood :   190–220 yds (174–201m)
5-Wood :   170–190 yds (155–174m)
7-Wood :   160–180 yds (146–165m)
3-Iron :   170–190 yds (155–174m)
4-Iron :   160–180 yds (146–163m)
5-Iron :   150–170 yds (137–155m)
6-Iron :   140–160 yds (128–146m)
7-Iron :   130–150 yds (119–137m)
8-Iron :   120–140 yds (110–128m)
9-Iron :   110–130 yds (101–119m)
Pitching :   90–110 yds (82–101m)
Sand Wedge: up to 80 yds (73m)

---

[2] https://www.golf-simulators.com/physics.htm.

4. **Puzzle Problem:** (after Taha)

> You have four chains each consisting of three solid links. You need to make a bracelet by connecting together all four chains. It costs 2 cents to break a link and 3 cents to resolder it.

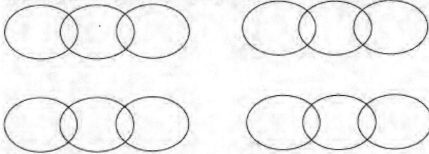

5. **TSP Tour creation problem:**

> Following Chapter 6 in David Wolber's book www.appinventor.org/book2 entitled Map Tour, please create this app on the computer. Display your Designer Screen and Blocks used to make your app and a possible screen shot of the emulator.
>
> > ▸ Extra credit: Create a map tour of another scenic place such as Boston, our Umass campus etc. your favorite vacation spot.
> > ▸ Look at the suggestions presented by Wolber for customizing a tour.
> > ▸ Could this be the basis of an optimization project?

2

Combinatorial Optimization $G(V, E)$

Overview Combinatorics is generally concerned with the arrangement, grouping, ordering, or selection of a discrete set of objects usually finite in number [10]. Combinatorial Optimization (CO) is concerned with finding the "best" arrangement, grouping, or ordering of a set of discrete objects. Combinatorial Optimization problems are a sub-class of OR and CS problems for which many applications abound. Graph algorithms such as the Minimum Spanning Tree and Shortest Paths are excellent example problems for CO. Figure 2.1 illustrates alternative shortest path topologies for an accessibility map at the University of Massachusetts to help guide people across the campus.

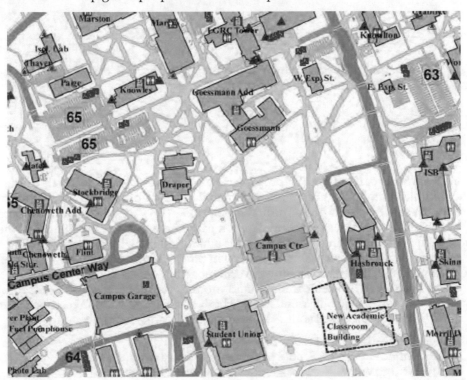

Fig. 2.1 University of Massachusetts Accessibility Map

Keywords: History, Matching, Enumeration, Graph Routing and Network Design Problems

© Springer Nature Switzerland AG 2021

J. MacGregor Smith, *Combinatorial, Linear, Integer and Nonlinear Optimization Apps*,
Springer Optimization and Its Applications 175,
https://doi.org/10.1007/978-3-030-75801-1_2

Proving as we have just done, the pigeon-hole principle by Linear Programming (LP) techniques may be reminiscent of the use of a sledge-hammer to crack the proverbial walnut.

—V. Chvatal

Perhaps even more than to the interaction between man and nature, graph theory is based on the interaction of human beings with each other.

—Dénis König

J.H.C. Whitehead, the great English topologist said of Dénis König (*a graph theorist*), '*He works in the slums of Topology*'.

—Paul Erdos

2.1 Introduction

There are many different types of CO problems which are normally framed in terms of a graph context $G(V,E)$ where G is the graph with a finite set of nodes V and edges E. There are sets of alternatives and we usually try to compare the objects in the sets until one of them emerges as the best solution. Occasionally, we enumerate all the objects in the sets to find the best alternative. A profit/cost function is usually supplied for comparing the different alternatives.

One particularly important CO problem is the Traveling Salesman Problem (TSP) discussed next. I start the first lecture of the MIE 379 course with this TSP example.

2.2 Stowe Cycle Traveling Salesman Problem (TSP)

The TSP is one of the most well-known and most important CO problems. Many people recognize its significance and universality.

2.2.1 Introduction

There are two major parts of the TSP problem. The first is to find a feasible tour through the set of nodes V also known as a Hamilton cycle and the second problem is to find the optimal Hamilton cycle of all possible ones. The first problem is $\mathcal{NP}$-Complete, a decision problem, while the second is $\mathcal{NP}$-Hard, the optimization part where, from all feasible tours, the best is selected. Thus, even finding a feasible tour is a hard problem, let alone optimizing it.

Let's say that we run a bicycle touring club in Stowe, Vermont, and we wish to plan a tour for our group members in the Stowe, Vermont area.

There are a number of scenic trails in and around the town, and you wish to plan a small tour of the town for your friends that is both challenging and interesting. Figure 2.2 represents a map of Stowe, Vermont . The four principal stops in our tour are

SV: Stowe Village, the main area of the town.
SC: The old school house on School street leading away from the village.
EB: Emily's Bridge which is a historic covered wooden bridge.
LV: The Lower Village which is a commercial area along Route 100.

What is of importance here is not necessarily the stops but the edges/arcs connecting them which reflect the beautiful views and landscape and the ever-changing levels in elevation that provide the challenge for the bike riders.

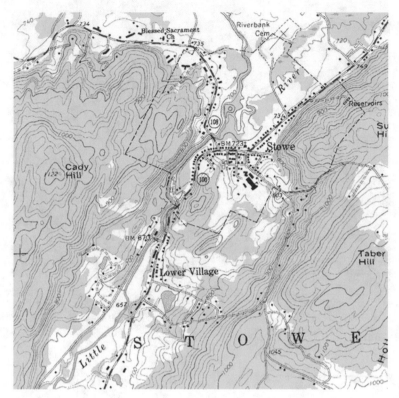

Fig. 2.2 USGS Map of Stowe, Vermont

2.2.2 Problem

In one sense, if we consider the arcs connecting the points of interest as the road distances between them, we seek the most interesting tour between them. Since there can be n possible tour sites, it can be shown that there are $(n-1)!$ possible tours. We need a way, an algorithm, to find the "best" route (Fig. 2.3).

If we think about the problem **carefully**, we see that because the tour is a cycle (no pun intended), it does not matter which tour stop we start with; it must always go through the starting stop, so we can eliminate one stop from the possible combinations, then only have to examine the remaining $(n-1)$ permutations. Figure 2.4 illustrates the conceptual ideas for a TSP solution .

Given this situation, we see that because there are only four stops on the tour, we could exhaustively enumerate all possible tours $(n-1) = 3! = 6$ and pick the best one. In general, this is **not** a good idea if n is larger $n \geq 10 \approx ((n-1) = 9!)$, but it is okay for smaller values and we can get an optimal solution. Enumeration of the tours makes the problem easily fit in the $\mathcal{N}\mathcal{P}$-Complete class of problems. So we are following a Backtracking strategy as discussed in Chapter 1.

Given a **preference matrix** connecting the tour stops, we can then evaluate the different possible tours. The values of the preference matrix are based on a scale from $1 \rightarrow 10$ where 10 is the *most interesting and the least stressing*, while 1 is the *most stressing and least interesting*.

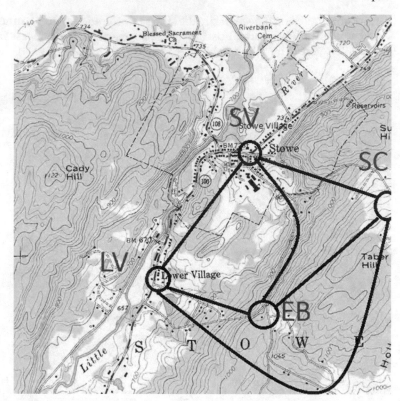

Fig. 2.3 G(V,E) Tour

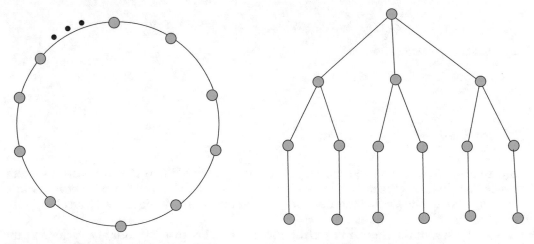

Fig. 2.4 a) Cyclic Nature of Problem (left)   b) Enumeration Tree (right)

If we factor into the problem such as the attribute *Safety* of the bicyclists, then our problem is most certainly challenging. These objectives actually are conflicting in nature so even our simple bicycle planning problem is a WP.

What other types of objective functions could be used? Distance, costs, time, profits, and quality come to mind.

Below is the preference matrix between the stops on the tour. Whose preferences are these and how do we measure them? We want to **maximize** our preference rather than minimize the distance between stops on the tour. Minimizing distance is normally the way people solve this TSP problem. Preferences are difficult to quantify in general, yet they allow for a more multi-objective type of problem. Do you prefer coffee with milk or without milk? How would you quantify the difference or would you be indifferent? If we place the preferences on a scale of *(lowest)* 1–10 *(highest)*, then it becomes a reasonable vehicle to compute the tour transitions.

$$
\begin{bmatrix}
 & 1(villlage(lv)) & 2(school(sc)) & 3(bridge(eb)) & 4(lower(lv)) \\
1(sv) & 0 & 6 & 8 & 3 \\
2(sc) & 2 & 0 & 5 & 9 \\
3(eb) & 6 & 3 & 0 & 6 \\
4(lv) & 4 & 4 & 5 & 0
\end{bmatrix}
$$

Below is the list of possible tours:

$$
\begin{array}{lll}
\text{a:} & 1 \rightarrow 2 \rightarrow 3 \rightarrow 4 \rightarrow 1 & \Sigma = 21 \\
\text{b:} & 1 \rightarrow 2 \rightarrow 4 \rightarrow 3 \rightarrow 1 & \Sigma = 26 \\
\text{c:} & 1 \rightarrow 3 \rightarrow 2 \rightarrow 4 \rightarrow 1 & \Sigma = 24 \\
\text{d:} & 1 \rightarrow 3 \rightarrow 4 \rightarrow 2 \rightarrow 1 & \Sigma = 20 \\
\text{e:} & 1 \rightarrow 4 \rightarrow 2 \rightarrow 3 \rightarrow 1 & \Sigma = 18 \\
\text{f:} & 1 \rightarrow 4 \rightarrow 3 \rightarrow 2 \rightarrow 1 & \Sigma = 13
\end{array}
$$

There are many possible objective functions here which might include minimizing vehicle exhaust pollution, uphill and downhill stress for the riders, and so on.

2.2.3 Mathematical Model

We want to achieve the optimal solution in a general situation. One mathematical formulation of the problem is given below, and it shows that if you could create this formulation one could solve it with a Backtrack-type algorithm, called Branch and Bound. Our formulation is based upon the one in Dantzig, Fulkerson, and Johnson [5].

M_a Mathematical Sets: (i,j) tour stops

P Parameters: n:– number of tour stops; p_{ij}:– cost/benefit or in our case the preferences to travel from stop i to stop j

D Decision variables: $x_{ij} = 1$ if we travel from stop i to stop j, and $x_{ij} = 0$ otherwise.

O Objective function:

$$
\text{Maximize } Z = \sum_{i=1} \sum_{j \neq i, j=1} p_{ij} x_{ij};
$$

C Constraints:

$$
\sum_{i=1, i \neq j} x_{ij} = 1 \; j = 1, \ldots, n \tag{2.1}
$$

$$
\sum_{j=1, j \neq i} x_{ij} = 1 \; i = 1, \ldots, n \tag{2.2}
$$

$$
\sum_{i \in Q} \sum_{j \neq i, j \in Q} x_{ij} \leq |Q| - 1 \; \forall |Q| \subseteq \{1, 2, \ldots, n\}, |Q| \geq 2 \tag{2.3}
$$

$$
x_{ij} \in \{0, 1\} \text{for all } i \text{ and } j \; : j > i \tag{2.4}
$$

The general mathematical model objective function looks to sum the preferences for all the possible tours, subject to the fact that each tour enters each node exactly once, Constraint equation (1), and visits all nodes exactly once, Constraint equation (2). One major difficulty with the problem is to avoid the subtours (Constraint equation #3) that can occur unless one is careful in the enumeration process.

2.2.4 Algorithm

The designer screen for AI2 is the primary vehicle for setting up the App representation. App Inventor 2 (AI2) as was discussed is basically comprised of three parts:

- **I. Designer Screen:** Input of images and parameter setups.
- **II. Visual Blocks Programming:** Logic, control, and procedural programming constructs.
- **III. Emulation or Phone App:** Feedback and Demonstration of App.

Figure 2.5 illustrates the designer screen for our App where all the parameters are established, along with the buttons, distance matrix, images, and screen inputs.

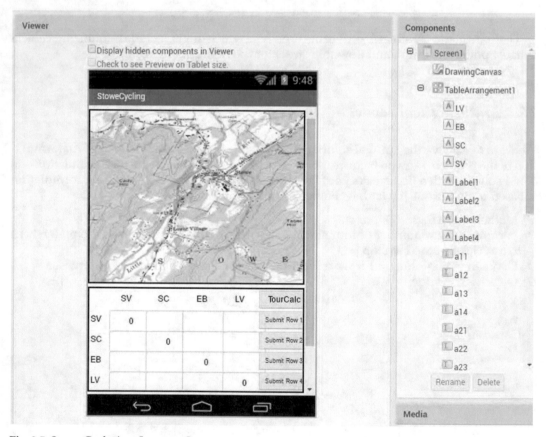

Fig. 2.5 Stowe Cycle App Inventor Screen

This TSP actually turns out to be a difficult combinatorial optimization to solve in general for a large number n of stops. Since the Stowe Cycle problem is very small, we can solve it with our enumeration algorithm implemented in AI2. Figure 2.6 is the set of blocks for the algorithm for the Stowe Cycle problem in AI2. It was pretty straightforward to program since only addition was required for the alternative routes.

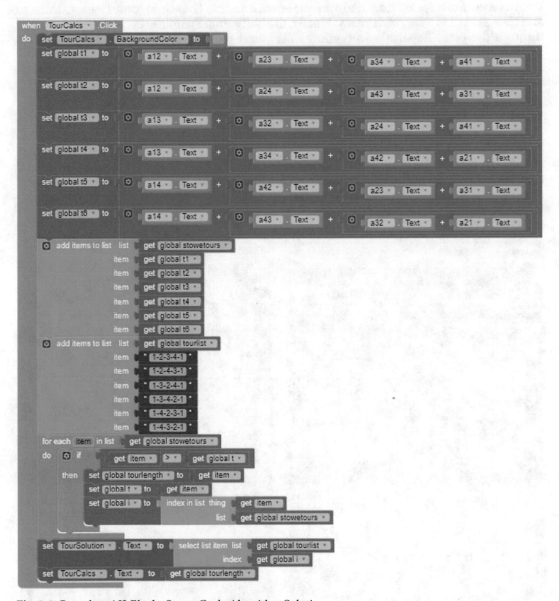

Fig. 2.6 Complete AI2 Blocks Stowe Cycle Algorithm Solution

The blocks algorithm is essentially in three steps:

- Compute the tour length of the Hamilton cycles.
- Create a list of the cycles.

- Compare the lists and choose the best one.

The list of data structures in AI2 is very important in most of the Apps.

In examining the blocks algorithm in Figure 2.6, the orange global variables for the tour lengths were first created from the input preference matrix. The actual routes were calculated and then added to a list, blue boxes. Finally, the individual Hamilton cycles were compared through the conditional blocks (yellow ones), and the largest value was identified and printed out on the screen (green blocks).

2.2.5 Demonstration

Here, in Figure 2.7, is a picture of the actual optimization App. On the left is the input screen and on the right is the tour calculation with $Z = 26$ at the top of the screen, and the optimal tour $1 - 2 - 4 - 3 - 1$ at the bottom of the screen. Finally, after inputting the sample data and running the Backtrack algorithm through the phone App, we have as might be expected

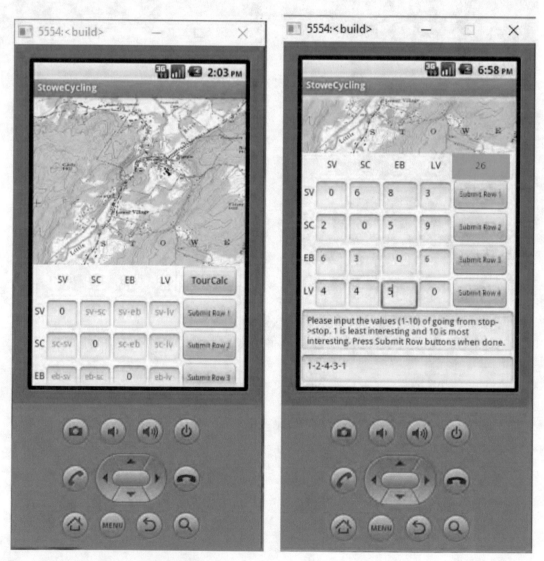

Fig. 2.7 Stowe Cycle App Solution

$$1 \rightarrow 2 \rightarrow 4 \rightarrow 3 \rightarrow 1 \qquad \sum = 26 \qquad\qquad (2.5)$$

2.2.6 Evaluation

The App works for a very limited number of tour stops, but it is designed to show the general structure of the backtracking strategy for combinatorial optimization problems. It is nice and compact for inputting the preference structure and the output is relatively fast for the problem.

Large problem instances where n is allowed to range larger will need to wait until the next section and in the Appendix where we can employ the NEOS server to solve larger problem instances.

Now let's examine a different problem concerned with warehouse management. It is somewhat related to the TSP problem, but can be treated from many different industrial engineering viewpoints, as can many of the WPs.

2.3 Warehouse Quick Pick

A very practical industrial engineering problem is concerned with an order picking system in a warehouse. As part of a supply/chain operation, order picking can be defined as the systematic activity of selecting a small number of goods from a warehouse system in order to satisfy a number of customer orders. We could set it up as a TSP, but we want to examine a different approach and show that even if we have a technique for solving a problem, formulating it from a different viewpoint will generate a different solution.

We will see that this App relies on the strategic principles of sorting and simple comparisons of the combinations to execute its logic. There can be many complex algorithmic and optimization procedures for this order picking problem as it involves location, time, the selection process, and the number of items to be picked.

2.3.1 Introduction

The University of Massachusetts has a warehouse of food items that need to be delivered to the various dining commons, snack bars, and cafes on campus every morning. Dining is 24/7. Connor Tremarche a student in the MIE 379 class in 2012 worked at the warehouse and in order to make his life easier, he came up with the idea of programming the warehouse order pick App.

2.3.2 Problem

Every morning, employees of the warehouse have 10–15 orders for which they must fulfill by selecting items to be delivered on pallets to the dining services. Because this is a process occurring every morning, a phone App can be very useful.

2.3.3 Mathematical Model

Connor identified five key performance variables to be included in the App through a preference system on a scale of 1–4 so he could deal with the different performance measure units (time, location, and number) (Fig. 2.8):

Performance Variable Definitions

A   Estimated delivery time (**Time: 4** being the shortest)
B   Location necessity (**Rank: 4** being the highest)
C   Number of locations (**Exact Number**)
D   Estimated pick time (**Time: 4** being the shortest)
E   Estimated number of pallets (**Exact number**)

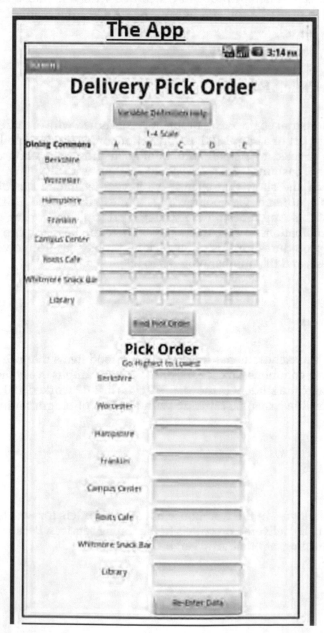

Fig. 2.8 Warehouse Quick Pick Input

These performance criteria were to be matched and the overall objective was to pick those items with the highest score. So in summary, we have the following preference ordering model where p_{ij} is the preference number of the performance variable and facility x_{ij} on the criterion:

$$\text{Maximize } Z = \sum_{i=1}^{5} \sum_{j \neq i, j=1}^{8} p_{ij} x_{ij}; \quad (2.6)$$

$$s.t. \ x_{ij} \in \{0,1\} \text{ for all } i \text{ and } j : j > i \quad (2.7)$$

2.3.4 Algorithm

Figure 2.9 illustrates a sample of the blocks programming used in the App. Simple arithmetical operations are used.

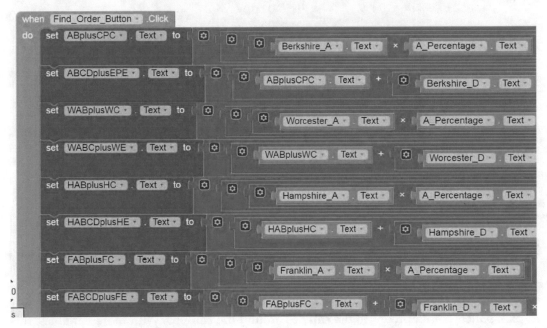

Fig. 2.9 Warehouse Quick Pick Blocks

2.3.5 Demonstration

As can be seen in Figure 2.10, this is a well-designed App. The problem instance has the five criteria on the top and the eight different facilities with cells below. Once the input data are entered, the App sorts the orders into a pick order working from highest to lowest. Sorting is the key algorithmic process.

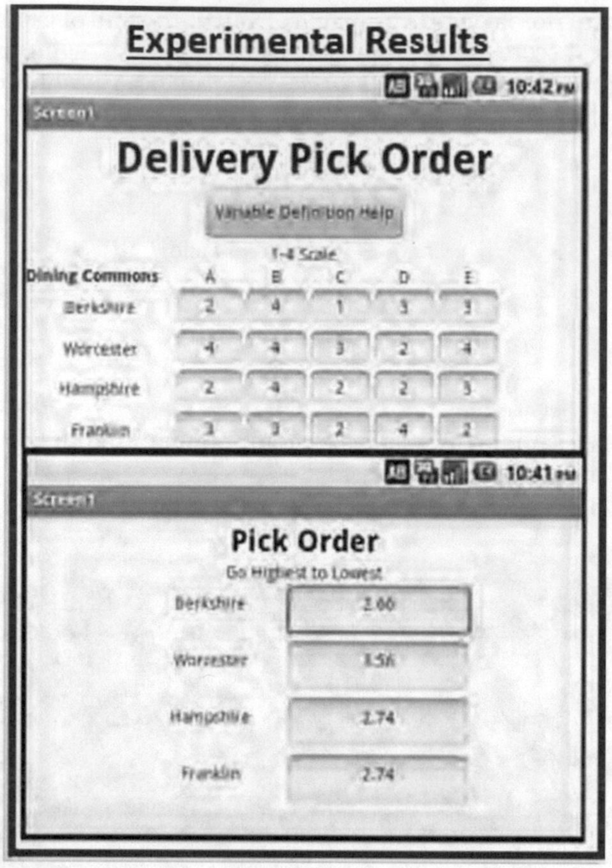

Fig. 2.10 Warehouse Quick Pick App

2.3.6 Evaluation

The goal of this App was to create an easy to read table with values of varying magnitudes. These outputs are ordered from highest to lowest in order to determine the optimal order to pick pallets for shipment to the eateries around the campus. With more variables, this App could become more accurate. Even with five variables, the calculated values closely reflect the day-to-day order that is currently used. This shows that a repetitious process can optimize itself with time.

2.4 Analytical Hierarchy Process (AHP)

Many times, we need to compare different alternatives and choose the best one, but we don't have a Linear Programming (LP) context in which to carry out the process. The context here can be considered as a decision-making problem under uncertainty rather than certainty in which LP occurs. So one procedure is to create a set of attributes/criteria to evaluate all the alternatives and use the criteria to select the best alternative. We do not *a priori* know the ranking of the criteria, so we must ask preference questions to structure the ranking of the criteria. This is the essence of the App. The Analytical Hierarchy Process (AHP) although sometimes controversial has become one effective way to rank criteria. It is discussed in Chapter fifteen of Taha's book [14]. We will provide a brief summary of the methodology.

2.4.1 Introduction

T.L. Saaty has developed an approach to ranking objectives and attributes along with alternatives that utilize the paired comparison approach in a very clever fashion. We will briefly summarize the methodology and illustrate with a small example. There are further details (*i.e.* consistency calculations) which I will not delve into but rather simply present.

Step 1.0: Construct a hierarchy of objectives/attributes and alternatives with as many levels as needed and a decision at the top. Figure 2.11 is an example of choosing a smartphone with two criteria/attributes and three alternatives smartphones.

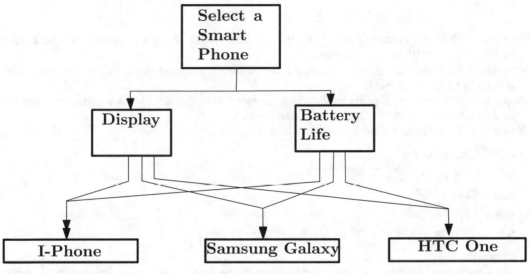

Fig. 2.11 Example Hierarchy

Step 2.0: For each level of the hierarchy, first for the attributes/objectives, then the alternatives, construct a pairwise comparison matrix and evaluate each cell pair a_{ij} in the matrix according to the following continuous scale (interpolation is also valid):

Value a_{ij}	Comparison description
1	Attribute i and j are equally important.
3	Attribute i is weakly more important than j.
5	Attribute i is strongly more important than j.
7	Attribute i is very strongly more important than j.
9	Attribute i is absolutely more important than j.

$$
\begin{array}{c c}
 & \begin{array}{c c c c c} a_1 & \quad a_2 & \quad \ldots & \quad a_r & \quad \Sigma_j \end{array} \\
\begin{array}{c} a_1 \\ a_2 \\ \vdots \\ a_r \\ \Sigma_i \end{array} &
\left[
\begin{array}{c c c c c}
a_{1,1} & a_{1,2} & \ldots & a_{1,r} & \Sigma_j \frac{a_{1j}}{r} = w_1 \\
a_{2,1} & a_{2,2} & \ldots & a_{2,r} & \Sigma_j \frac{a_{2j}}{r} = w_2 \\
\vdots & \vdots & \ddots & \vdots & \vdots \\
a_{r,1} & a_{r,2} & \ldots & a_{r,r} & \Sigma_j \frac{a_{rj}}{r} = w_r \\
\Sigma_i a_{i1} & \Sigma_i a_{i2} & \ldots & \Sigma_i a_{ir} & 1
\end{array}
\right]
\end{array}
$$

Step 3.0: Find the column sums $\Sigma_i a_{ij} \ \forall \ j$ columns which is used to normalize the values of the criteria within the columns.

Step 4.0: Find the row sums $\Sigma_j a_{ij}/r \ \forall \ i$ rows which computes the weights for each alternative $a_i \ \forall i$ across the criteria.

Step 5.0: Do this for all pairs in the hierarchy and sum the final values across the criteria and alternatives. Everything should add up to 1.

Step 6.0: Consistency Check

    6.1: Compute $\mathbf{Aw^t}$

    6.2: Compute $\frac{1}{r} \Sigma_{i=1}^{r} \frac{\text{ith entry in } \mathbf{Aw^t}}{\text{ith entry in } \mathbf{w^t}} = r_{max}$

    6.3: $CI = \frac{r_{max} - r}{r - 1}$

    6.4: $RI = \frac{1.98(r-2)}{r}$

    6.5: $CR = \frac{CI}{RI}$, If $CR < .10$ ranking is consistent.

Example

Let's say we wish to buy a new smartphone such as A: Apple iPhone, B:Samsung Galaxy, and C:HTC One. Let's further argue that two objectives/criteria are most important to us:

- Screen Display (D): pixel resolution, color quality, etc./
- Battery Life (B): # of hours, ease of re-charging, etc.

$$
P =
\begin{array}{c c}
 & \begin{array}{c c c} D & \quad B & \quad \Sigma_j \end{array} \\
\begin{array}{c} D \\ B \\ \Sigma_i \end{array} &
\left[
\begin{array}{c c c}
1 & 1/5 & \\
5 & 1 & \\
6 & 1.2 &
\end{array}
\right]
\end{array}
$$

$$
P =
\begin{array}{c c}
 & \begin{array}{c c c} D & \quad B & \quad \Sigma_j \end{array} \\
\begin{array}{c} D \\ B \\ \Sigma_i \end{array} &
\left[
\begin{array}{c c c}
.17 & .17 & .17 \\
.83 & .83 & .83 \\
1.00 & 1.00 & 1.00
\end{array}
\right]
\end{array}
$$

So through our analysis, battery life is more important than the display screen. Just to illustrate another example and its ranking, now let's add a third attribute: Web Navigation (N): Wi-Fi capability, Bluetooth, etc. The first matrix represents the scores for the three different attributes. The second matrix represents the calculations of the weights of the attributes by the AHP.

$$P_1 = \begin{array}{c} \\ D \\ N \\ B \\ \Sigma_i \end{array} \begin{array}{cccc} D & N & B & \Sigma_j \\ \left[\begin{array}{ccc} 1 & 3 & 5 \\ 1/3 & 1 & 3 \\ 1/5 & 1/3 & 1 \\ 23/15 & 13/3 & 9 \end{array} \right] \end{array}$$

$$P_2 = \begin{array}{c} \\ D \\ N \\ B \\ \Sigma_i \end{array} \begin{array}{ccccc} D & N & B & \Sigma_j \\ \left[\begin{array}{cccc} 15/23 & 9/13 & 5/9 & 0.63341 \\ 5/23 & 3/13 & 1/3 & 0.26045 \\ 3/23 & 1/13 & 1/9 & 0.10614 \\ 1 & 1 & 1 & 1.00 \end{array} \right] \end{array}$$

In this example demonstration, the ranking is reversed. Let's check the consistency of our new ranking. Carrying out the consistency check to give reassurance to our ranking of the criteria for buying the smartphone, we carry out Step 6.0.

Step 6.0:   Consistency Check

6.1:   Compute $\mathbf{Aw^t}$

$$\begin{pmatrix} 1 & 3 & 5 \\ 1/3 & 1 & 3 \\ 1/5 & 1/3 & 1 \end{pmatrix} \begin{pmatrix} .63341 \\ .26045 \\ .10614 \end{pmatrix} = \begin{pmatrix} 1.97546 \\ 0.8000 \\ 0.3256 \end{pmatrix}$$

6.2:   Compute $\frac{1}{r}\sum_{i=1}^{r} \frac{\text{ith entry in} \mathbf{Aw^t}}{\text{ith entry in } \mathbf{w^t}} = r_{max} = 3.101$

6.3: $CI = \frac{r_{max}-r}{r-1} = 0.05053$

6.4: $RI = \frac{1.98(r-2)}{r} = 0.66$

6.5: $CR = \frac{CI}{RI}$, If $CR < .10 = 0.0766 < 0.10$ ranking is consistent

So, we are pretty comfortable with our current criteria ranking. For further details about the AHP, please see Taha's Chapter 15.

2.4.2 Marching Band AHP App

Now let's see how the AHP is used to rank order the members of the marching band. In the marching band at the University of Massachusetts, there are a fixed number of marching spots available. Occasionally, there are more candidates than drill spots so that the section leaders need to have a systematic procedure to choose the candidates.

2.4.3 Problem

Amanda Skriloff a student in 2016 in the MIE 379 course and in marching band designed the App. She identified five performance criteria essential to the selection problem.

- [Playing Ability (PA):] Ability of a person to play scales, rhythm, and show music.
- [Marching Ability (MA):] Ability of a person to march in proper form and execution.
- [Class Conflicts (CC):] Missing rehearsals due to class conflicts is critical. Because the marching band meets every day, being at rehearsal is important.
- [Improvement Potential (IP):] Candidate's ability to improve and grow with practice.
- [Enthusiasm (E):] Overall enthusiasm of a person. How passionate and enthusiastic is the candidate.

2.4.4 Mathematical Model

Pairwise comparisons are made for each of the attributes. There are $\frac{(n^2-n)}{2}$ or ten comparisons.

For Amanda's problem, she used a 5×5 paired comparison matrix that is used to compare and rank-order the criteria. The $\square$ is one of the ten items to be answered in the App.

$$
\begin{array}{c}
\\ PA \\ MA \\ CC \\ IP \\ E
\end{array}
\begin{array}{c}
PA \quad MA \quad CC \quad IP \quad E \\
\left[
\begin{array}{ccccc}
-- & \square & \square & \square & \square \\
 & -- & \square & \square & \square \\
 & & -- & \square & \square \\
 & & & -- & \square \\
 & & & & --
\end{array}
\right]
\end{array}
\tag{2.8}
$$

The ten total comparisons are made according to the following rules:

- If both attributes are equally important, a 1 is placed.
- If the attribute on the RIGHT is more important, a number greater than 1 (3, 6, or 9) is placed meaning that the item on the right is x times more important than the left item.
- If the attribute on the LEFT is more important, a number less than one (1/3, 1/6, or 1/9) is placed meaning that the item on the left is y times more important than the right item.
- The matrix is then squared and there is a summation of each row and a grand summation of each row total. Each row total is then divided by the grand total. The highest number represents the most important attribute. The lowest number is the least important attribute.

2.4.5 Algorithm

Figure 2.12 illustrates a sample of the blocks used for programming the App.

2.4.6 Demonstration

As can be seen in Figure 2.13, the App works very well and is nicely designed.

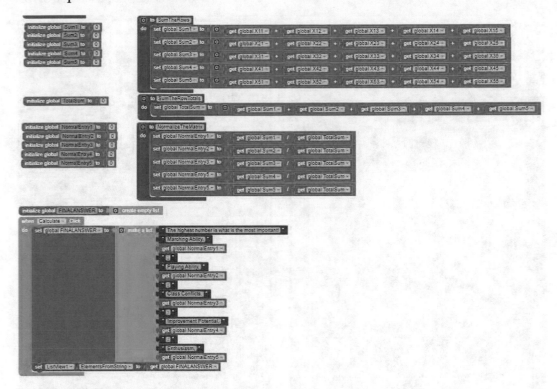

Fig. 2.12 AHP Algorithm Blocks Programming

2.4.7 Evaluation

The App uses the methodology of the AHP to create a rank ordering of the criteria. It does not actually rank the candidates which is a downside, but the problem of coming up with the ranking of the criteria is a challenging first step to this process. It is a well-designed App. Let's examine a recipe finder with the AHP process.

2.5 Recipe Selection Problem

The purpose of this App is to select the best recipe from a predetermined list based on user inputs. The user will be asked to compare three criteria (cost, nutritional value, and cook time) using the Analytical Hierarchy Process (AHP) ranking system. Based on predetermined local priorities for nine available recipes, the App will use the user's inputs to select the recipe with the highest global priority in the AHP model.

2.5.1 Introduction

The AHP methodology is actually very general as will be shown in this current App for recipe selection and may also be inferred from the previous App. All the actions encountered by AHP will be reflected in the properties and characteristics of the stand-alone App, and no NEOS server is required. D. Solkol programmed this App.

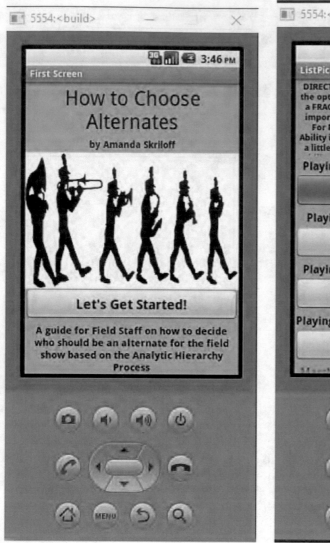
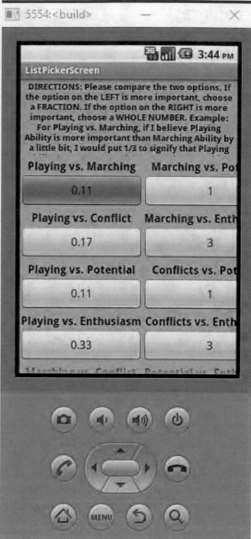

Fig. 2.13 AHP AI2 Output Apps

2.5.2 Problem

The AHP allows users to make complex decisions by giving each alternative a priority (or weight) based on the user's preferences. Using the values below, comparisons are made between the different criteria themselves (for this App, these are the user inputs) and then between the alternatives in terms of each criteria (these are programmed into the App). Using the AHP methodology, a global priority is found for each alternative and a decision can be easily made by choosing the alternative with the highest value (Fig. 2.14).

- User inputs;
- Determine the importance of each criteria;
- Determine priorities for all alternatives for each criteria;
- Calculate global priorities and select the best option.

Comparison Values to be entered in the matrices:

1/9	B is absolutely more important than A	3	A is weakly more important than B
1/7	B is very strongly more important than A	5	A is strongly more important than B
1/5	B is strongly more important than A	7	A is very strongly more important than B
1/3	B is weakly more important than A	9	A is absolutely more important than B
1	A and B are equally important		

 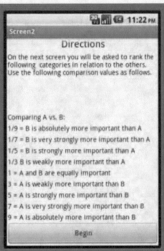 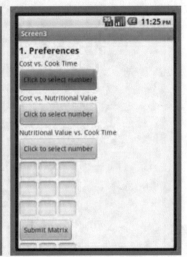

Fig. 2.14 Recipe Inputs/Outputs

2.5.3 Mathematical Model

By using quantitative metrics for each category (cost = $/lb. of main ingredient; nutritional value = calories; time = minutes), each pair of recipes can be compared based on how close they are in relation to the other. To calculate the comparison values, it was assumed that a low cost, low calorie, quickly prepared meal is best. In the first line of the second table, the recipe 1 is 7 times better than recipe 2 because its cook time is 35 minutes shorter. Similar comparisons are made for all alternatives for each criterion. Below are the values for each criterion: cost, nutritional value, and time (Table 2.1).

Recipe	Cost	Nutritional Value	Time
1	0.17	0.02	0.20
2	0.15	0.08	0.05
3	0.17	0.05	0.03
4	0.08	0.13	0.22
5	0.17	0.27	0.12
6	0.17	0.13	0.02
7	0.02	0.19	0.22
8	0.06	0.09	0.13
9	0.02	0.04	0.02

Table 2.1 Data Inputs

From these comparison values, a normalized matrix is created using the AHP method. The resulting weights for each alternative are called local priorities (Fig. 2.15).

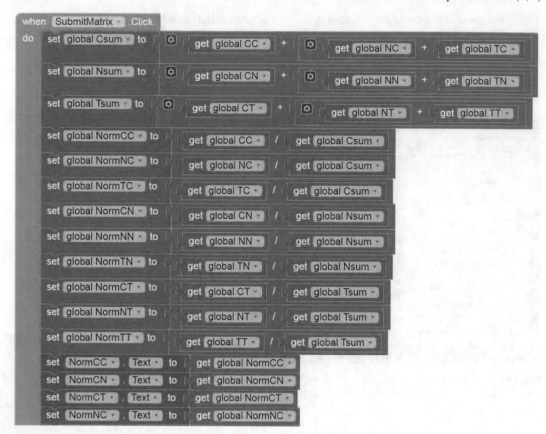

Fig. 2.15 Normalized Matrix Calculation

2.5.4 Algorithm

Using the weights from the user input and the local priorities, global priorities are now determined. The global priority sum for each alternative is the $\sum (local\ priority) * (criteria\ weight)$. The recipe with the highest global priority sum will represent the best option for the user (which is recipe #5) (Fig. 2.16 and Table 2.2).

	Cost		Nutritional Value		Cook Time		
	Local	Global	Local	Global	Local	Global	GLOBAL SUM
1	0.17*0.18	0.0301	0.02*0.75	0.0129	0.20*0.7	0.0142	0.0572
2	0.15*0.18	0.0275	0.08*0.75	0.0583	0.05*0.7	0.0033	0.0891
3	0.17*0.18	0.0301	0.05*0.75	0.0399	0.03*0.7	0.0019	0.0720
4	0.08*0.18	0.0146	0.13*0.75	0.0939	0.22*0.7	0.0159	0.1244
5	0.17*0.18	0.0301	0.27*0.75	0.2038	0.12*0.7	0.0084	**0.2423**←
6	0.17*0.18	0.0301	0.13*0.75	0.0960	0.02*0.7	0.0012	0.1273
7	0.02*0.18	0.0030	0.19*0.75	0.1457	0.22*0.7	0.0159	0.1646
8	0.06*0.18	0.0110	0.09*0.75	0.0639	0.13*0.7	0.0095	0.0844
9	0.02*0.18	0.0038	0.04*0.75	0.0337	0.02*0.7	0.0012	0.0386

Table 2.2 Priorities

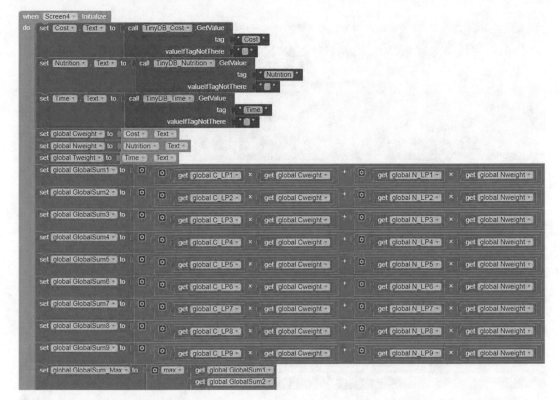

Fig. 2.16 Recipe Consistency Calculation

2.5.5 Solution App

Figures 2.17 and 2.18 illustrates the example solution process.

2.5.6 Evaluation

The AHP methodology is suitable for many different kinds of problems. For the purposes of this App, it works quite well. As is shown in this section, the App successfully takes the user's input preferences and uses the AHP methodology to deliver the best recipe option. MIT App Inventor 2 does pose challenges when it comes to the aesthetics and speed of the App. For the purposes of this task, however, the program is well suited.

Now, let's describe some different applications, one having to do with selecting a snowboard, then one for a scheduling problem, and the others concerned with tree network design problems.

2.6 Snowboard Selection

Jared Marvel designed this App because when he was first introduced to snowboarding, he had no idea what kind of board to choose. There are many snowboards on the market today, and the process of picking a snowboard can be very confusing. This App is meant to help snowboarders of all sizes, genders, and athletic abilities to select the right board (Figs. 2.19 and 2.20).

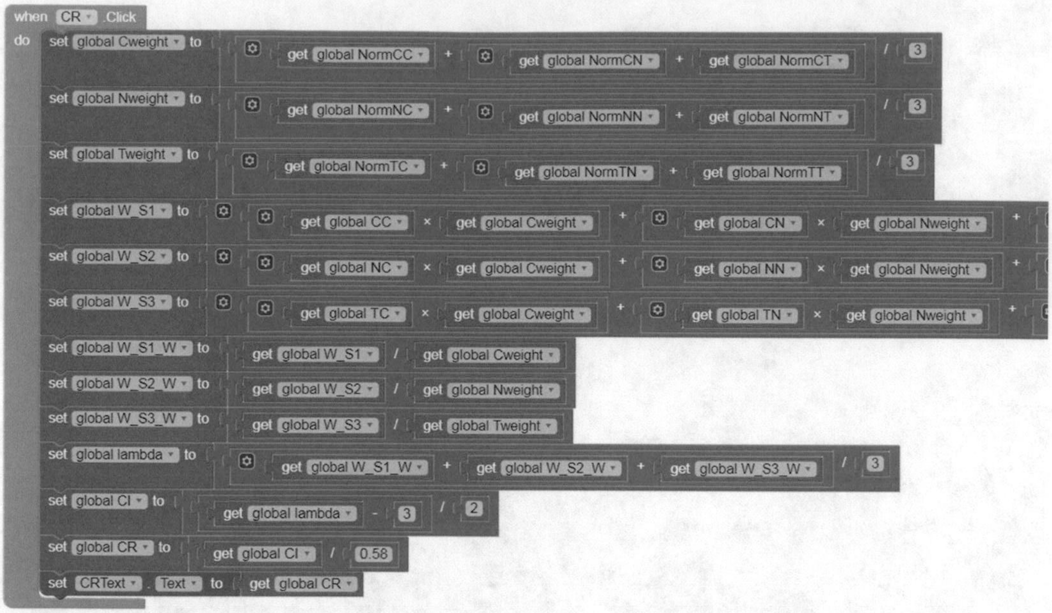

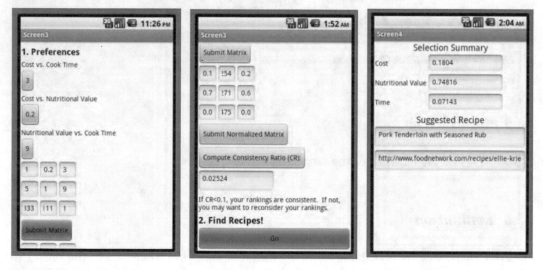

Fig. 2.17 Recipe AHP Blocks Scoring Calculation

Fig. 2.18 Recipe Outputs

2.6.1 Introduction

For the following App, we demonstrate the use of the "ListPicker" command in App Inventor which is a useful construct in programming for a smartphone as it helps structure the data input process. Snowboard Assistant is designed to take the inputs of the user's weight, gender, boot size, ability, and style and output the optimal snowboard parameters of length, waist width (width at the point where the bindings are attached), and board flex. This will assist the user in picking his or her new snowboard.

2.6.2 Problem

Figure 2.19 illustrates the input screen of the App.

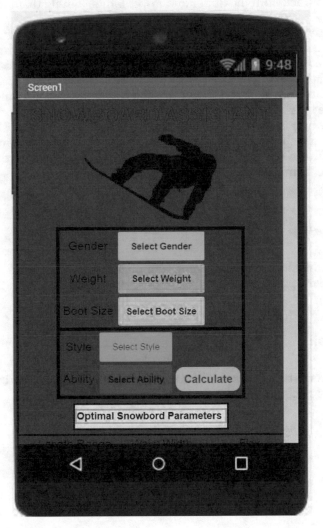

Fig. 2.19 Snowboard Input

2.6.3 Mathematical Model

There is no mathematical model for this App. Rather, it relies on the logical structuring of the parameters through a question–answer process.

2.6.4 Algorithm

The algorithm is essentially complete enumeration. Since most snowboard length charts use weight ranges and waist width recommendations along with boot size ranges, it made more sense to use list pickers instead of text boxes. This also makes the actual calculation easier, since we could use *if and elseif* statements to decide the length ranges.

Since certain options such as *freestyle* for the style list picker affect the range differently (they may decrease the upper limit, or increase the lower limit), Jared decided to make global variables to track both the upper and lower limits, joining them together at the end to create one single output for the range of lengths.

Since the recommendations for waist width are very similar, only different by the size range (because the sizing of men's boots and women's boots are different), he decided to make an *if statement* that would change the options of the boot size list picker depending on whether the user is male or female.

The stiffness output is simply decided by which style of riding is selected by the user. For instance, if the user selects freestyle, the output is soft. Most snowboard companies run their stiffness by a number ranging 1–10 (1 being soft and 10 being stiff). The reason he did not give a specific number for stiffness is because most snowboarders have their particular preference for stiffness, and this was provided to give a new snowboarder the basic range for them to choose from.

2.6.5 Solution App

A presentation of the blocks and their programming is shown in Figure 2.20.

For the snowboard problem, also, an example run is shown in Figure 2.21.

2.6.6 Evaluation

While the App is very straightforward and easy to use, it is self-contained and very efficient. Now let's examine a scheduling problem.

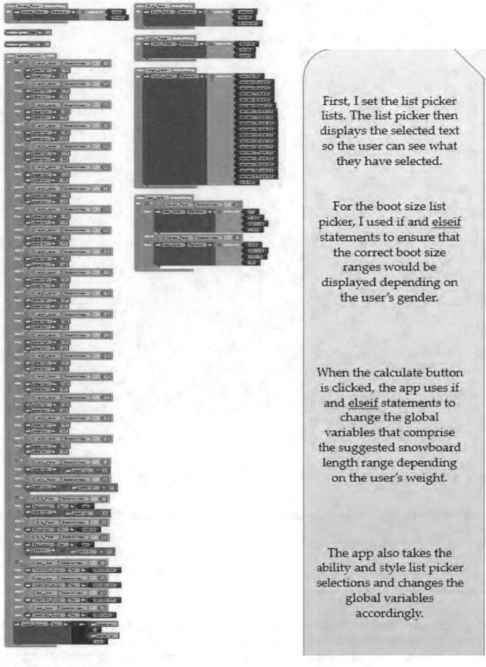

First, I set the list picker lists. The list picker then displays the selected text so the user can see what they have selected.

For the boot size list picker, I used if and elseif statements to ensure that the correct boot size ranges would be displayed depending on the user's gender.

When the calculate button is clicked, the app uses if and elseif statements to change the global variables that comprise the suggested snowboard length range depending on the user's weight.

The app also takes the ability and style list picker selections and changes the global variables accordingly.

Fig. 2.20 Snowboard Blocks

2.7 ABC Machine Flow Shop Scheduling Problem

Probably, one of the most elegant algorithms developed for an optimization problem is S.M. Johnson's two-machine flow shop scheduling algorithm [8]. While seemingly a trivial problem, it is a truly elegant algorithm and deserves a special place in algorithm design.

Example

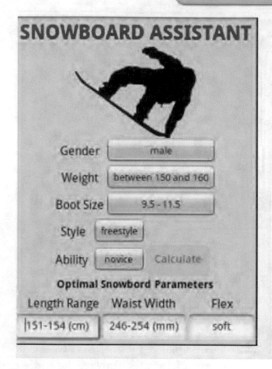

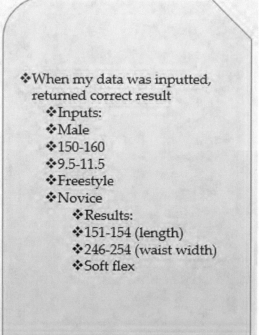

Fig. 2.21 Snowboard Output

2.7.1 Introduction

This problem is an excellent one to learn how to work with lists and procedures in AI2. It is a fundamental scheduling problem.

2.7.2 Problem

There are n items/jobs which must traverse one production stage machine and then another one. There is only one machine at each stage and at most, one item can be on a machine at a given time. Jobs are passed to each machine in the same order, once the list of jobs is found. The total number of machines is m. An example of a Gantt chart for a flow shop of five jobs on $m = 3$ machines is indicated in Figure 2.22. The schedule for the five jobs is as given and we seek to find the schedule to minimize the overall makespan (*i.e.* total time to complete all the jobs) .

The processing times p_i of the machines are given as input vectors $A_i, B_j, i, j = 1, 2, \ldots n$ which are positive values but otherwise arbitrary. The A_i include the setup time and processing time on the ith machine and correspondingly on the second machine. We seek the optimal scheduling order of items in order to minimize the total elapsed time of the jobs. Since the jobs proceed sequentially on the first and then the second machine, a solution schedule is a permutation vector $c = (1, 2, \ldots n)$ of the jobs on the machines, thus, the name of the algorithm *ABC*.

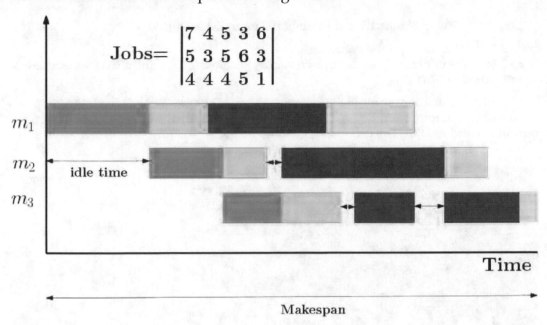

$$\text{Jobs=} \begin{vmatrix} 7 & 4 & 5 & 3 & 6 \\ 5 & 3 & 5 & 6 & 3 \\ 4 & 4 & 4 & 5 & 1 \end{vmatrix}$$

Fig. 2.22 Gantt Chart for a Flow Shop

2.7.3 Mathematical Model

We assume that the time for each job is constant and job times are mutually exclusive of the job sequence. An obvious extension of the problem is to consider stochastic processing times, yet there is no simple algorithm for the stochastic flow shop problem. All jobs must be processed in the first machine before going on to the second machine and there are no priorities.

The overall *makespan* of the two machines and their jobs is given by the following objective function value, where S is the optimal sequence of jobs:

$$F(S) = \max_{1 \leq i \leq n} \{p_i(S)\} \tag{2.9}$$

Even the *makespan* calculation is nonlinear, so the problem is quite difficult.

2.7.4 Algorithm

The algorithm is basically applied dynamic programming and is solvable in polynomial time for $m = 2$, but the problem is NP-Complete for $m \geq 3$ which is perhaps surprising. Under certain conditions for $m = 3$, it will produce an optimal sequence, but in general it will not yield the optimal solution.

The algorithm works directly with lists.

Step 1.0:   Create two lists A, B with the processing times of the jobs on the two machines. Create one output list, initially empty. Set the indices $k = 1, j = 0$.

   Step 1.1:   If the minimum time of all the jobs is on the first machine, place it in the output list C_1 at the beginning of the list. Set $k = k + 1$.
   Step 1.2:   If the minimum job is from the second machine, place it in the output list C_2 at the end of the list $n - j$, set $j = j + 1$.
   Step 1.3:   Remove or delete the job from both input machine lists.

Step 1.4:   Repeat Steps 1.1 and 1.2 until the input list is empty.

Step 2.0: C has the optimal sequence S.

Step 3.0:   The minimum makespan is created by generating the Gantt chart for the schedule embodied in $C(S)$.

Figure 2.23 illustrates some of the key procedural AI2 blocks for the scheduling App. We need to find the minimum for each of the processing lists, then sequence those jobs on the output list and readjust the processing times as we go.

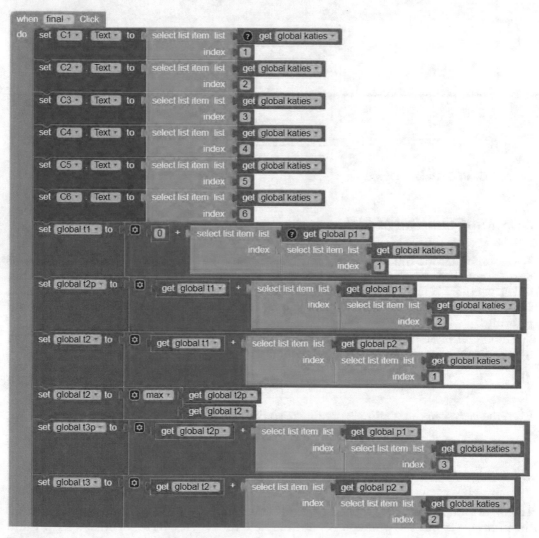

Fig. 2.23 ABC Scheduling Flow Shop Blocks

2.7.5 Demonstration

Figure 2.24 illustrates the App for an example flow shop problem with the processing times, optimal schedule, and time makespan for the schedule.

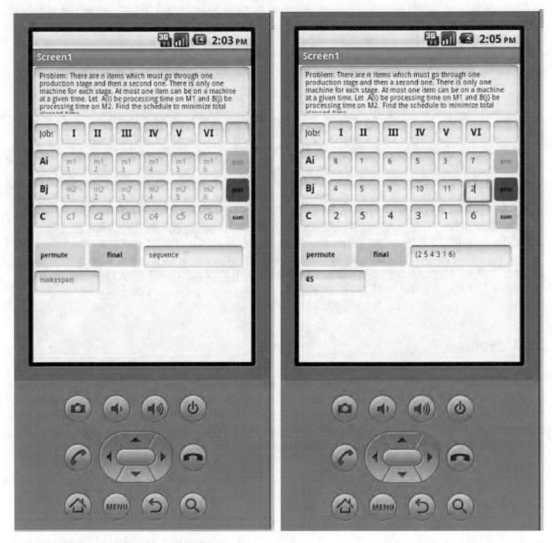

Fig. 2.24 ABC Scheduling Flow Shop Example

2.7.6 Evaluation

The App works pretty well. It is a good demonstration of list manipulation and the design of procedures for AI2.

2.8 Triage Scheduler

Ian Taylor a student of MIE 379 in 2017 programmed this App for evaluating patients coming into a hospital.

Everyday, hospitals are faced with the challenge of triaging, the sorting of and allocation of treatment to patients according to a system of priorities designed to maximize the number of survivors.

2.8.1 Introduction

In a perfect world, there will always be additional space and healthcare professionals readily available to tend to each patient, however, that is not the reality of the situation.

TriageSolver is an App that makes this reality much fairer and easier to manage. Intended for use by triage nurses, the user's first step is to input the total number of rooms in a given department floor (Ex: Intensive Care Unit (ICU)) as well as the number of occupied rooms. The user then continues to enter quantitative information for each patient in the queue (Ex: Glasgow coma scale, Systolic blood pressure, Respiratory Rate). TriageSolver then calculates a score for each of the patients based on several parameters, constraints, and user inputs, and returns a list of their scores. TriageSolver optimizes patient care and hospital efficiency by telling the user which patients to admit based on the severity of their condition. In the future, TraigeSolver or other applications like it will replace traditional triaging methods and greatly improve the healthcare industry.

2.8.2 Problem

After the user enters all of the data for a patient as well as room availability, the user will then click Triage! and TriageSolver will calculate and record that patient's score. This is done by creating an objective function within the App that maximizes patient's scores in order to determine who is in the most critical condition. When data is entered into the system, the App treats these values as decision variables which are displayed below.

2.8.3 Mathematical Model

$x_1 =$ Glasgow coma scale
$x_2 =$ Systolic blood pressure (mmHg)
$x_3 =$ Respiratory rate (breathes/min)
$x_4 =$ Open or depressed skull fracture
$x_5 =$ Two or more proximal long-bone fractures

$x_6 =$ Amputation proximal to wrist or ankle
$x_7 =$ Paralysis
$x_8 =$ Chest wall instability or deformity
$x_9 =$ Pregnant (20 weeks or more)
$x_{10} =$ Length of stay one week or more

2.8.4 Algorithm

The decision variables must follow a series of parameters and constraints in order for (1) the App to accept the inputs and (2) to calculate the score for a given patient. The constraints

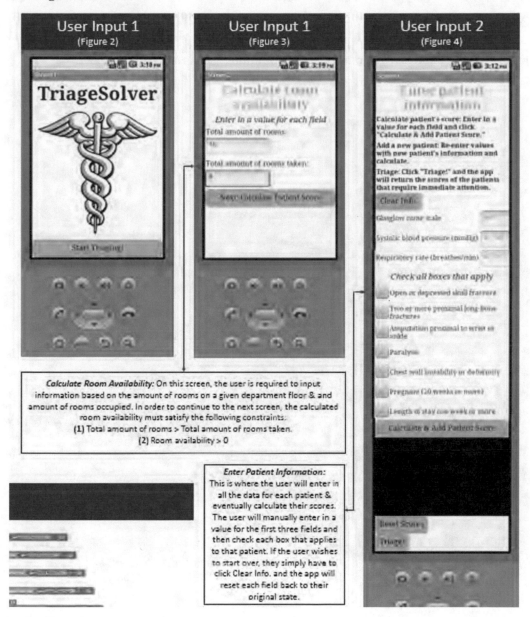

Fig. 2.25 Triage Inputs

are modeled after the standard triage guidelines for assessing a patient's condition. For example, if a patient's systolic blood pressure is recorded at 190 mmHg or more, the App assigns more weight to decision variable x_2 in order to indicate that patient condition is quite severe; of course, their overall score depends on the rest of the decision variables as well. Once the user is done inputting data in for each patient, TriageSolver will indicate which patients require immediate attention by retrieving only the highest scores from the patient database. The number of patients selected will also satisfy the room availability constraint (Fig. 2.25).

2.8.5 Demonstration

Figure 2.26 illustrates a sample of the programming blocks of this App.

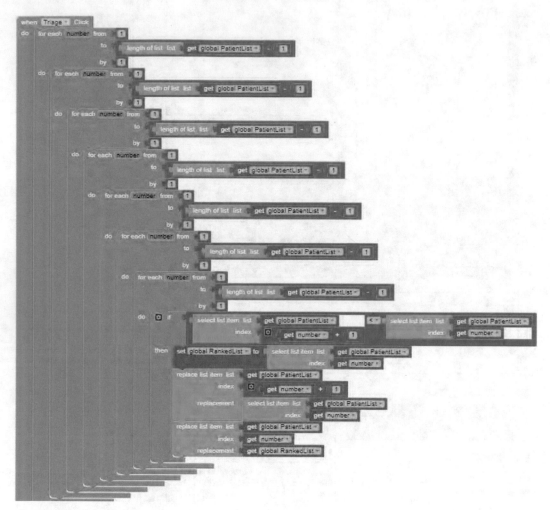

Fig. 2.26 Triage Blocks

2.8.6 Evaluation

Using MIT App Inventor 2, Ian was able to design an App called TriageSolver that can be used by any healthcare professional. The challenge of triaging in a hospital is a hectic and serious task, that is why TriageSolver is designed to automatically prioritize patients in the queue based on their assessed condition. TriageSolver optimizes hospital efficiency and maximizes the survival rate by identifying which patients need immediate care (Fig. 2.27).

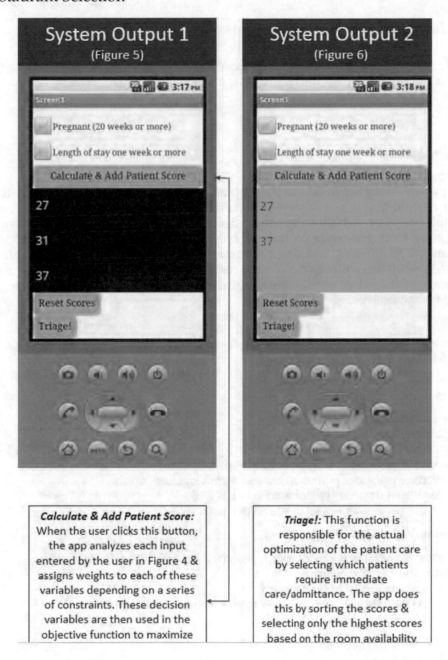

Fig. 2.27 Triage Solution

2.9 Restaurant Selection

Avery Stroman (2017) programmed this restaurant selection App for the area around the University of Massachusetts campus. Avery wanted to make this process more efficient. She took the two most important factors, food type and price, and created an App that helps its user find the best eatery for them.

2.9.1 Introduction

Many factors influence where people choose to go out to eat. Whether it comes from a friend's recommendation or the restaurant is conveniently located, finding the best place to grab food depends on the person looking.

2.9.2 Problem

Assumptions: Avery assumed that the App user was only interested in dining at the most popular restaurant types in Amherst, which include Chinese, Italian, Mexican, American, Pizza, and Breakfast. Additionally, she limited the amount of restaurants that were within each type of food. For example, there is only information about five different Mexican restaurants, five different Breakfast restaurants, etc. within the App. More assumptions were made when determining the average cost per person at each restaurant. She also took the average of the prices of the entree, figuring that every customer would typically buy just one for themselves.

2.9.3 Mathematical Model

Notation

- $i=$ type of food, $i = 1, 2, \ldots 6$
- 1 = Chinese, 2 = Italian, 3 = Pizza
- 4 = Mexican, 5 = American, 6 = Breakfast$j$=j the option of food type $ij = 1, 2, \ldots 5$
- x_{ij}= name of restaurant of type of food type i, option j
- p_{ij}= average price per person to eat restaurant x_{ij} User Inputs
- s_i= selection of desired type of food $s_i = 1, 2, \ldots 6$
- d = desired price user wants to spend per person (≥ 0)

   Decision Variables

- x_{ij}=name of restaurant j from user's preferred type of food i
- **Constraints:** $u_{ij} = |(d - p_{ij})|u_{ij} \geq 0$
- x_{ij} and p_{ij} correspond to each other $i = s$
- **Objective Function:** *Minimize u_{ij}, report back corresponding x_{ij}*

2.9.4 Algorithm

The algorithm searches for all the restaurant alternatives that satisfy the user's constraints so it is an exhaustive search of all the combinations.

2.9.5 Solution App

Figure 2.28 illustrates the input information for the App, while Figure 2.29 shows the programming blocks used in the App.

Figure 2.29 shows some of the blocks for the Chinese restaurants as well as other blocks for choosing different types of restaurants.

The screenshot on the left is the App Inventor Design Screen. The center picture is taken from the Phone Emulator and shows the screen that pops up when the user clicks "Choose Food Type" allowing them to select from the six different types of food. The right image shows the optimal results when the user wants to spend around 11 dollars per person on Chinese food.

Fig. 2.28 Restaurant Screen App Input

2.9.6 Evaluation

In conclusion, this App could be very useful for both college students on a budget and local Amherst residents. The App is very user-friendly and a reliable resource. The user gets a recommendation from the application almost instantaneously, which in this day and age is almost expected from technology. Moreover, Avery imagined the information it is providing to be extremely useful to those using it. Additionally, the structure of the code is flexible, meaning it can be updated as prices and restaurants change over the years.

2.10 Minimum Spanning Tree Network Design Problems

Another classic combinatorial optimization problem concerns the construction of Minimum Spanning Trees (MSTs). These are problems appropriate for a Greedy Solution methodology.

2.10.1 Introduction

We shall examine Kruskal's algorithm. In Taha in Chapter 6, he presents Prim's algorithm. Both are greedy approaches and Kruskal's relies on sorting the edges of the graph to struc-

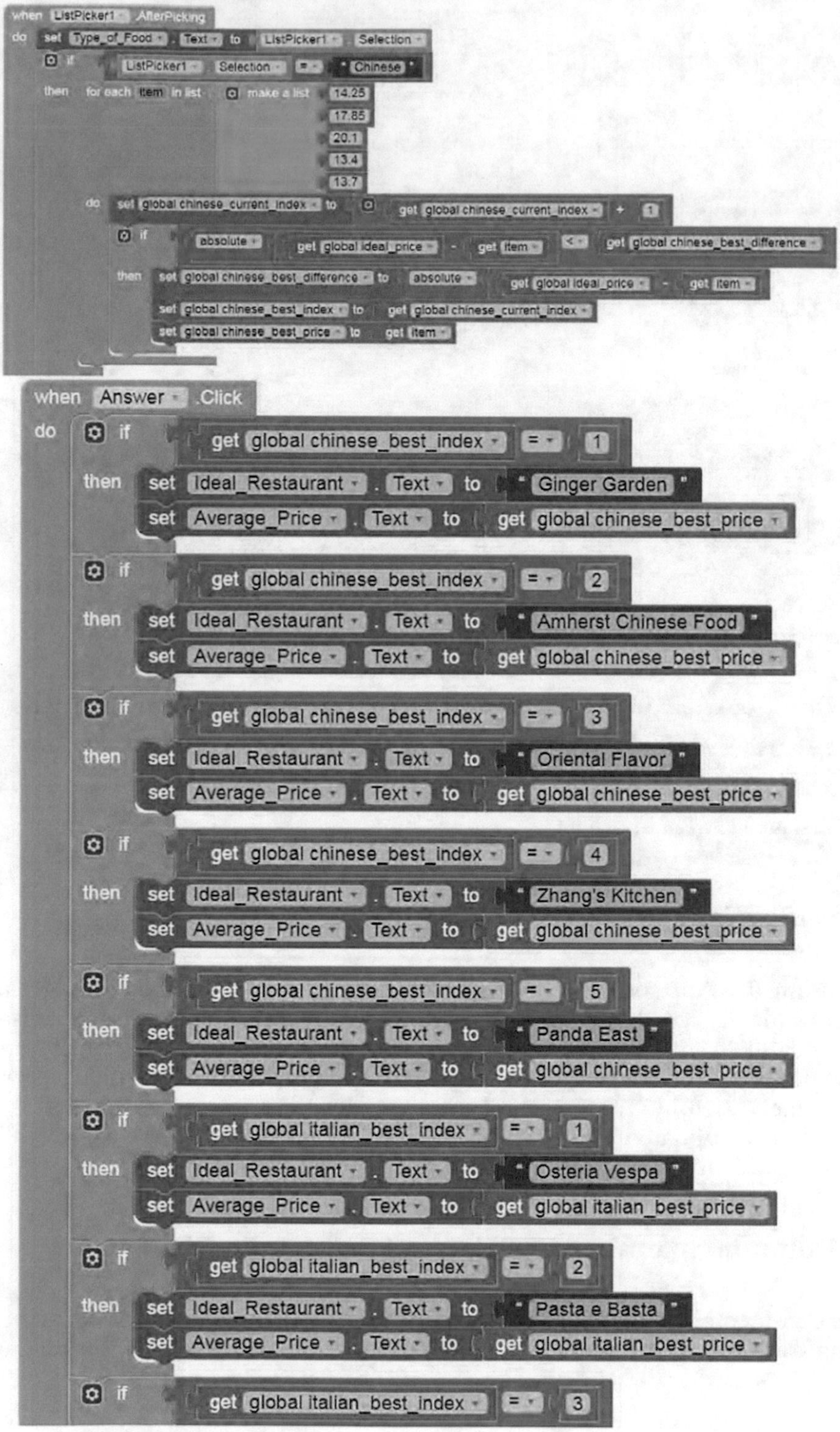

Fig. 2.29 Restaurant Blocks

ture a solution. This is the reason why we chose Kruskal's algorithm, since it relies on the procedure *sorting*.

There is even another algorithm attributed to Boruvka (1920s) who famously toured Europe lecturing with his algorithm. In essence, his algorithm is also similar to a parallel programming approach. We are given a graph and we wish to connect all the nodes together. An *acyclic graph* is one that consists of no *cycles*, *i.e.* graphs that loop back on themselves. A connected acyclic graph is called a *tree*. In a tree T, any two vertices (nodes) are connected by exactly one *path*. A *subtree* of a graph is a subgraph which is a tree. If the tree is a spanning subgraph, it is called a *spanning tree*. We are interested in finding the *minimum* spanning tee.

2.10.2 Problem

For an example, a telephone company wishes to rent a subset of existing cables each of which connects two cities (or locations within cities). The rented cables should suffice to connect all cities and they should be as cheap as possible. In solving this problem, we shall use a Minimum Spanning Tree (MST) **T** which spans N nodes such that

$$\sum_{(i,j)\in T} c_{ij} \text{ is a minimum} \tag{2.10}$$

There are many applications of MSTs.

o highway maintenance	o cable tv networks
o pipeline distribution	o HVAC systems in buildings
o long distance telephone rates	o airline routes
o statistical clustering	o molecular modeling

2.10.3 Mathematical Model

[**def:**] Given a Graph **G(N,A)** that is an undirected connected graph. A subgraph $T' = (N, A')$ of G is a spanning tree of G if and only if T is a tree and connects all nodes in N. If each arc has a weight (cost) c_{ij}, then we seek to minimize the overall construction cost of the tree T. It is probably more direct to program this via CO rather than LP.

2.10.4 Algorithm

We shall follow Kruskal's algorithm for constructing the MST which is probably the easiest to understand and probably the best one to execute if you are doing it by hand. This step is a classic example where the sorting of the edge data sets up the greedy approach for building the tree (Fig. 2.30).

Kruskal's Algorithm:

Step 1.0:   Sort the edges of G in increasing order by cost.
Step 2.0:   Keep a subgraph $S \in G$, initially empty.
Step 3.0:   For each edge $e \in G$ in sorted order
   If the endpoints of e are disconnected in S
   add e to S.
Step 4.0:   return S.

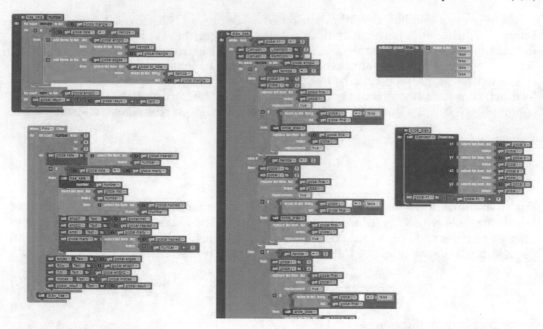

Fig. 2.30 MST Screen Apps

Whenever we add an edge (u,v), it is always the smallest connecting the part of the subgraph with the rest of G without cycles, so it is always a part of the MST. The worst-case complexity of the algorithm is where the sorting of the edges occurs, so this requires $O(N\log n)$ time in the worst case. The rest of the steps of the algorithm are bounded by the running time of the sorting step.

2.10.5 Demonstration

Figure 2.31 illustrates the input screen of the App which requires the user to input points on a map. In this case, the U.S.G.S. map is of San Francisco, California. The App will compute the distance matrix between the points, then sort the distances so that the MST algorithm parallels Kruskal's approach for finding the MST.

2.10.6 Evaluation

In Figure 2.32, we see the generation of the sorted edge list and the list of the edges which are part of the MST solution. This is for an example problem input through the canvas map of the City of San Francisco. The algorithm is very fast, but of course this is a small problem. In the worst case, the algorithm is bounded by the creation of the distance matrix which is $\mathcal{O}(\mathcal{N})^2$. Still, it is very effective and the simple graphic output is quite revealing of the MST topology.

Fig. 2.31 MST Screen App Input and Output

2.11 Piping Design and Analysis

This problem was inspired by Ryan Barnes along with Alex Barth and Anthony Broding who put together this interesting engineering application in 2011. Ryan was working for an engineering firm dealing with this problem. Ryan subsequently graduated and continued working with the engineering firm after graduation.

2.11.1 Introduction

This App is a sophisticated engineering analysis program for designing and repairing underground pipes, their probability of failure, and the location of the pipe problem.

2.11.2 Problem

In underground pipe network design, one must account for the internal pipe pressures, vertical earth loads, depth of pipe, diameter, deflection and stress, and service source, as

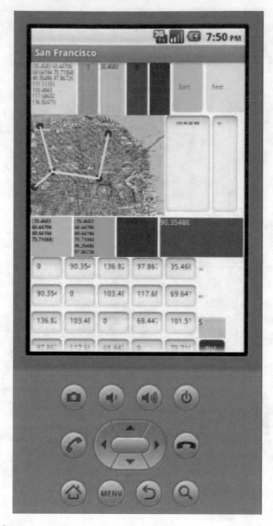

Fig. 2.32 MST Graphical Output

well as numerous other design factors. How all these are combined is the subject of this App.

2.11.3 Mathematical Model

The mathematical relationships depend upon the data surrounding the piping breakdown and the customers' service requirements. There are many separate equations necessary given the data parameters and these must all be integrated within the App (Figs. 2.33 and 2.34).

2.11.4 Algorithm

The algorithm is designed to gather all the data and enumerate the solutions to calculate the probability of failure. Figures of 2.35 illustrate this input process.

Figure 2.36 illustrates the programming blocks.

Fig. 2.33 Pipe Inputs

Figure 2.37 shows the solution of the process and the map of the location of where the problem occurs .

2.11.5 Solution App

Because of the many inputs, the Probability of Failure (POF) is determined by many more factors than could be analyzed by hand. This App would be very useful in which planning projects should be carried out by the company.

2.11.6 Evaluation

The App is very sophisticated and the last link for locating the pipe on Google Maps is impressive. The next problem is a benchmark Combinatorial Optimization (CO) problem, the Shortest Path problem, which occurs in many applications.

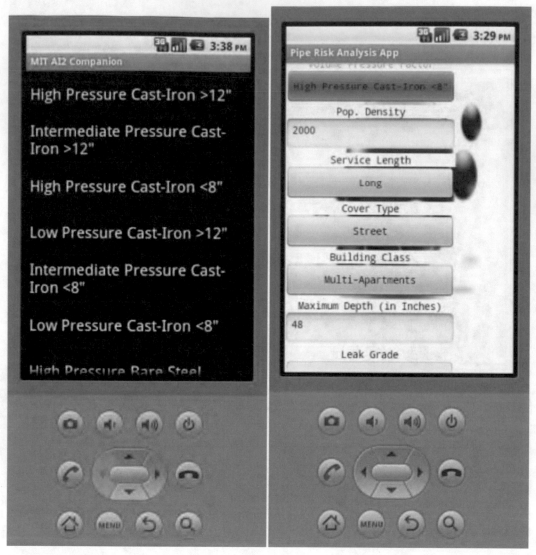

Fig. 2.34 Pipe Inputs

2.12 Shortest Path Problems

Shortest Path problems are classic Combinatorial Optimization problems and are eminently suitable for a smartphone environment. They represent problems appropriate for a recursive Dynamic Programming (DP) approach since the "stages" of the DP approach are often considered the nodes of the $G(V,E)$ and the "states" are the remaining distance information functions at each node. We will illustrate Dijkstra's algorithm which is a classic approach to the problem and is the algorithm used in Google to search for the shortest path in Google Maps.

2.12.1 Introduction

The history of the SPT is rather recent computationally (1950's–60's), although historically finding shortest paths is deemed rather a rudimentary problem for most species.

Fig. 2.35 Pipe Inputs

1. What we would like to find is that given a specified pair of nodes (i, j), determine the shortest path between i and j.
2. Secondly, we might want to find the shortest path tree between i and all other nodes in the graph, or even the set of shortest paths between all pairs on nodes in the graph.
3. Finally, we might want to find the shortest chains between all pairs of nodes in the graph.

- Figure 2.38 is that of a unit length dodecahedron mapped onto the plane. So the edge weights are all equal to 1.
- The figure next to it is the shortest path tree on the graph assuming we are starting at vertex #1 which is at the top of the graph and we wish to connect to all the nodes in the graph.

Example Applications include

a)   Manufacturing Process Planning
b)   Equipment Replacement
c)   Longest Paths (CPM/PERT Networks)
d)   Most Reliable Routes
e)   Bottleneck Arc Problems

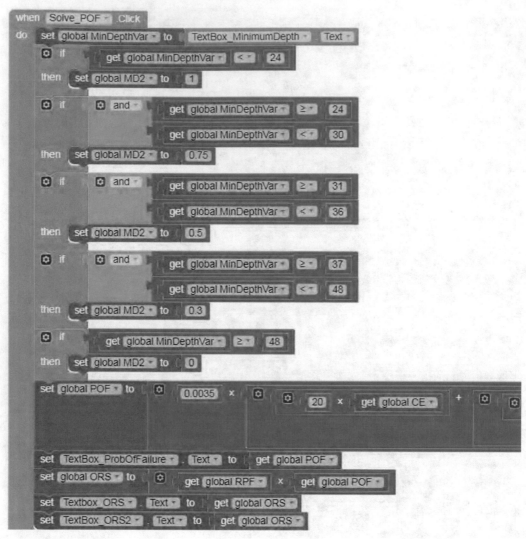

Fig. 2.36 Pipe Blocks

2.12.2 Problem

We are given a graph $G(V,E)$ with costs on the arcs and where in the simplest case, we wish to find a path of forward arcs leading from the source node to a destination sink node.

In certain stochastic reliability situations, the arc costs may represent probabilities, but we will not address that issue here. There are some complications with detecting negative cycles, so we assume that all the arc costs are nonnegative.

2.12.3 Mathematical Model

The following mathematical model is a graph-oriented version of the problem.

[**Problem:**] Given an $G(V,E)$ with cost/weight/distance c_{ij} associated with each arc $a_i \in A$, find the shortest (minimal) chain between two specified nodes s and t in $G(V,E)$.

i) In general, we have the following assumptions which may be applicable to most any type of shortest path application as we shall illustrate:

Fig. 2.37 Pipe Outputs

ii) c_{ij} can be negative (certain application may allow for negative arcs, *e.g.* profits)

iii) $c_{ij} = \infty$ if no arc exists between i and j.

iv) Negative circuits are not allowed.

v) $c_{ik} + c_{kj} \leq c_{ij}$ (a.k.a. the triangle inequality).

2.12.4 Algorithm

We will describe Dijkstra's algorithm (1959). It is one of the most famous algorithms for solving shortest path problems.

Problem: Given a $G(N,A)$ with weight c_{ij} associated with each arc $a_i \in A$, find the shortest (minimal) chain between two specified nodes s and t in $G(N,A)$.

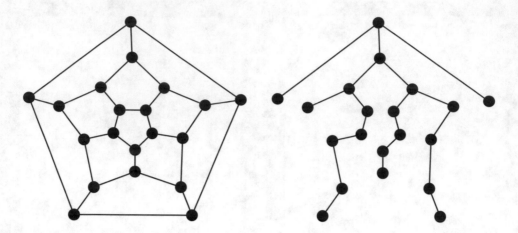

Fig. 2.38 Dodecahedron Graph Shortest Path

i) In general, we have the following assumptions which may be applicable to any type of application as we shall illustrate:

ii) c_{ij} can be negative (certain applications may allow for negative arcs, *e.g.* profits)

iii) $c_{ij} = \infty$ if no arc exists between i and j

iv) negative circuits are not allowed.

v) $c_{ik} + c_{kj} \leq c_{ij}$ (a.k.a. the triangle inequality).

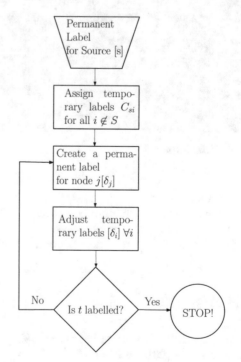

Define two sets of nodes

$\mathscr{L} :=$ nodes already labeled

$\bar{\mathscr{L}} :=$ unlabeled nodes

$A :=$ arcs in the shortest path tree

$\delta_j :=$ a labeling function on each $x_j \in N$

$\delta_j :=$ a permanent label

which represents an actual shortest chain to node x_j

Step 0.0: $[\delta_s] = 0$; $\delta_i = c_{si}$; $c_{si} = \infty$ (if no arc)

Set iteration counter $k = 1$

Step 1.0: $\delta_{last} = \min_{i \in \mathscr{L}} \delta_i$

$j = last :=$ the last node to get a permanent label

$\bar{\mathscr{L}} :=$ set of nodes with temporary labels.

Step 1.1: $[\delta_j] \leftarrow \delta_{last}$ and $\mathscr{L} = \mathscr{L} \cup x_j$; $A = A \cup a_{ij}$

Step 1.2:   If $[\delta_t]$ is found terminate.

Step 2.0: For each $x_i \in \bar{\mathcal{L}}$ replace δ_i with

$$\delta_i = \min_{i \in \mathcal{L}}\{\text{old } \delta_i; [\delta_{last}] + c_{ji}\}$$

Step 2.1: $k \longleftarrow k+1$
Step 2.2:   Return to Step 1.0.

- $$\sum_{k=1}^{N-1} 3(N-k) = 3[(N-1) + (N-2) + (N-3) + \ldots + (N-N-1)]$$

- $$\frac{3N(N-1)}{2} \Rightarrow O(N^2)$$

Figure 2.39 illustrates an example problem for the algorithm. We shall follow the steps of the algorithm for its solution.

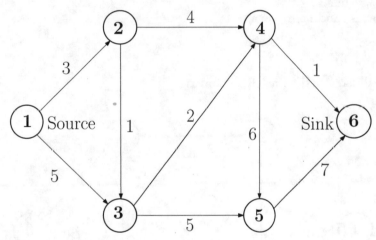

Fig. 2.39 Example for SPT Problem

- Step 0.0: Set $[\delta_1] = 0$ and $\delta_i = c_{si}$ where possible; $k = 1$
- Step 1.0: $\delta_{last} = 3$ so $j = x_2$

  Step 1.1: $[\delta_2] = 3; \mathcal{L} = \mathcal{L} \cup x_2$ and $A = A \cup a_{12}$
  Step 1.2: $[\delta_6]$ not found.

- Step 2.0: $\delta_i = \min_{i \in \bar{\mathcal{L}}}\{\text{old } \delta_i; [\delta_{last}] + c_{ji}\}$
- $\delta_3 = \min\{\delta_3; \delta_2 + c_{23}\} = \{5; 3+1\} = 4$
- $\delta_4 = \min\{\delta_4; \delta_2 + c_{24}\} = \{\infty; 3+4\} = 7$
- $\delta_5 = \min\{\delta_5; \delta_2 + c_{25}\} = \{\infty; 3+\infty\} = \infty$
- $\delta_6 = \min\{\delta_6; \delta_2 + c_{26}\} = \{\infty; 3+\infty\} = \infty$

  Step 2.1: $k \longleftarrow 2$
  Step 2.2:   Return to 1.0.

- Step 1.0: $\delta_{last} = \min_{i \in \mathcal{L}}\{\delta_i\}$
- so $\delta_{last} = 4$ corresponding to $j = x_3$

  Step 1.1: $[\delta_3] = 4; \mathcal{L} = \mathcal{L} \cup x_3$ and $A = A \cup a_{23}$
  Step 1.2: $[\delta_6]$ not found.

- Step 2.0: $\delta_i = \min_{i \in \bar{\mathcal{L}}}\{\text{old } \delta_i; [\delta_{last}] + c_{ji}\}$
- $\delta_4 = \min\{7; 4+2\} = 6$

- $\delta_5 = \min\{\infty; 4+5\} = 9$
- $\delta_6 = \min\{\infty; 4+\infty\} = \infty$

  Step 2.1: $k \longleftarrow 3$
  Step 2.2:   Return to 1.0.

- Step 1.0: $\delta_{last} = \min_{i \in \mathscr{L}}\{\delta_i\}$
- so $\delta_{last} = 6$ corresponding to $j = x_4$

  Step 1.1: $[\delta_4] = 6; \mathscr{L} = \mathscr{L} \cup x_4$ and $A = A \cup a_{34}$
  Step 1.2: $[\delta_6]$ not found.

- Step 2.0: $\delta_i = \min_{i \in \mathscr{L}}\{\text{old } \delta_i; [\delta_{last}] + c_{ji}\}$
- $\delta_5 = \min\{9; 6+6\} = 9$
- $\delta_6 = \min\{\infty; 6+1\} = 7$

  Step 2.1: $k \longleftarrow 3$
  Step 2.2:   Return to 1.0.

- Step 1.0: $\delta_{last} = \min_{i \in \mathscr{L}}\{\delta_i\}$
- so $\delta_j = 7$ corresponding to $j = x_6$

  Step 1.1: $[\delta_6] = 7; \mathscr{L} = \mathscr{L} \cup x_6$ and $A = A \cup a_{46}$
  Step 1.2: $[\delta_6]$ found.

Figure 2.40 illustrates the final SPT solution.

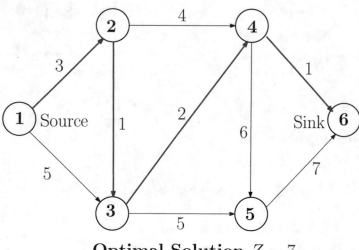

Optimal Solution $Z = 7$

Fig. 2.40 **Optimal Solution for SPT Problem**

Also, the simplex algorithm of the NEOS server can solve our problem, so the LP setup is pretty standard.

2.12.5 *Google Maps Demonstration*

As a demonstration of the SPT algorithm, let's employ Google Maps and build upon their excellent map graphics and implementation of Dijkstra's algorithm. This App was developed by Sydney Hauver, a mathematics major, in the 2017 MIE 379 class and he was able to build upon the Google software to solve some very practical SPT problems (Fig. 2.41).

Fig. 2.41 Input Screens for SPT Problem

Figure 2.42 shows the Google map of Amherst, Massachusetts, with another screenshot of a downtown restaurant Johnny's for which we want to find the shortest time path from our residence. The resulting optimal path was non-intuitive and interesting to the user.

Figure 2.43 illustrates the blocks for the App. Sydney has been able to integrate AI2 with Google Maps through blocks programming. He has put together the restaurants, gas stations, and hotels in and around the UMass Amherst campus along with the current location of the user to generate the shortest paths from the user's source location to any one of a set of destinations. The map graphics are spectacular.

2.12.6 Evaluation

One can execute this App in the context of your phone location for your own particular location, so it is a very general implementation. The App is pretty straightforward and works quite well. In the next section, we illustrate a practical application of the shortest path methodology where we also employ Dijkstra's algorithm.

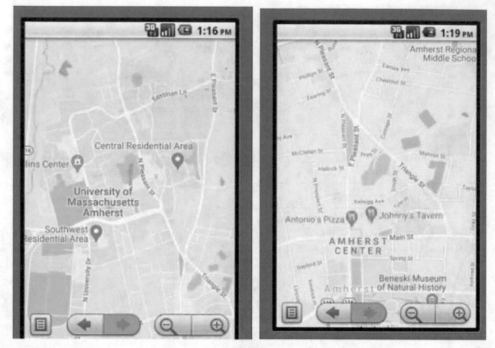

Fig. 2.42 Input Screens for Google Maps SPT Problem

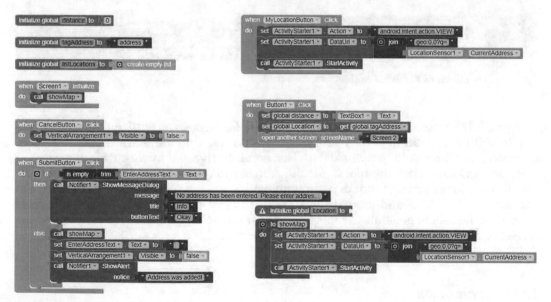

Fig. 2.43 Blocks for Google SPT Problem

2.13 Shortest Path Pump Out

This App was programmed in the first App class in 2012 and since this is the first class in which we attempted to use AI2, we had little experience in programming any sophisticated algorithms or use of the NEOS server. The algorithm is self-contained.

2.13.1 Introduction

This is an application of the shortest path methodology to fulfill orders for maritime gasoline, along with some side constraints. The student worked at a marina during the summers and knew the problem quite well.

2.13.2 Problem

We are given a set of four wharf/marina locations along the water in Rhode Island which dispenses gasoline to various numbers of boats at each location. Figure 2.44 shows the water body with boats in the water and is indicative of the problem environment. The users call the dispatch on their smartphones to request the supply boat to fulfill their fuel needs. So this is a very nice use of AI2 and the shortest path methodology.

Fig. 2.44 Pumpout Input and Solution

2.13.3 Mathematical Model

The mathematical model is based on the shortest path technology and is a clever use of integrating the technology and the smartphone.

2.13.4 Algorithm

The algorithm is fairly straightforward as Figure 2.45 demonstrates.

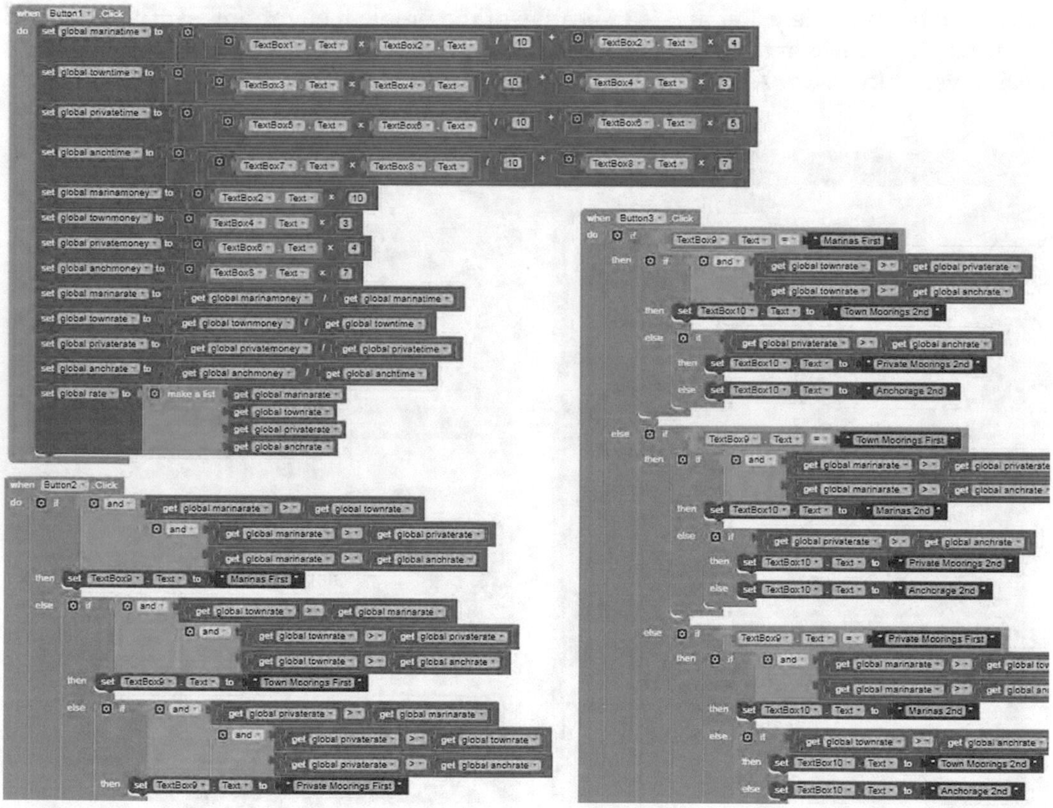

Fig. 2.45 Pumpout Blocks

2.13.5 Solution App

Figure 2.44 illustrates the solution which is generated very quickly.

2.13.6 Evaluation

This is a nicely designed App which is engineered in response to a very practical problem.

2.14 Matching and Linear Assignment Algorithm

The penultimate problem we shall examine in this section is the matching and linear assignment algorithm. It is a fundamental combinatorial optimization problem with a rich history. It is fairly simple to state yet it has a deep history with many contributors, properties, and many interesting approaches for a solution, including Linear Programming. The problem originates with the mathematician G. Monge 1784 who was interested in moving earth from one location to another.

2.14.1 Introduction

In its fundamental form, this problem is concerned with matching two sets of vertices $X \times Y$ in a graph with the smallest set of connecting edges with no common vertices in order to minimize an objective function. So there can be flows in the network, blossoms, paths, and trees to consider. In its most practical sense, it is concerned with matching pairs of potential spouses (*i.e.* the marriage problem) or matching many personnel with possible jobs (*i.e.* job assignment), so it can be a very useful construct. Let's say that we have a set of $x_1, x_2, \ldots, x_n$ workers for a set of n available jobs $y_1, y_2, \ldots, y_n$, $X \times Y$. Can we assign all the workers so that each worker has a job and each job has assigned only one worker? This is called the personnel assignment problem.

2.14.2 Problem

We will restrict ourselves to bipartite graphs as this characterizes the linear assignment problem for most optimization applications. A *matching* in a graph is an assignment of edges to the vertices in a graph such that no two edges share a common vertex. A maximal matching is a subset of edges such that no additional edges can be added to the graph to improve the value of the matching (Fig. 2.46).

In the language of graph theory, we have a bipartite graph with vertex set

$$V = X \bigcup Y \qquad \text{and edge set} \tag{2.11}$$

Each edge e connects a vertex of X to a vertex of Y. Moreover $|X| = |Y|$, which is important in making assignments. For each edge $e \in E$, we are given a nonnegative weight w_e. We want to find a subset $M \subseteq E$ of edges such that each vertex of X and Y is incident to exactly one edge of M and the sum $\sum_{e \in M}$ is a minimum.

- [Example] Let's say that we wish to minimize the material handling costs of all products flowing between some existing machines and some new machines we wish to locate in a facility. We want to *assign* each new machine to each available location. For our situation, let's look at an example layout. Squares represent present machines while pentagons represent candidate locations.

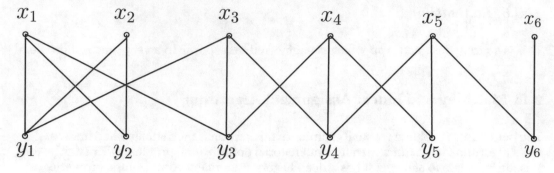

x_1x_2x_3x_4x_5x_6y_1y_2y_3y_4y_5y_6

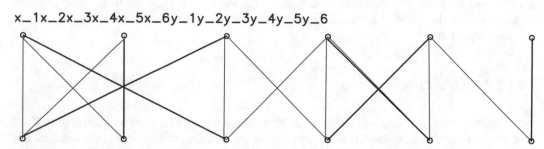

Fig. 2.46 Bipartite Graph (top) with Matching Assignment (bottom)

- The forklifts move the material between the machine centers on the plant aisles. How should we assign the new machines to the locations in order to minimize the material handling costs?

$$
\begin{array}{c}
\begin{array}{ccccc} 1 & 2 & 3 & 4 & 5 \end{array} \\
\begin{array}{c} M \\ C \\ P \\ N \end{array}
\begin{bmatrix}
10 & 12 & 4 & 8 & 6 \\
3 & 5 & 10 & 12 & 4 \\
7 & 3 & 5 & 4 & 8 \\
2 & 1 & 3 & 3 & 3
\end{bmatrix}
\end{array}
$$

- The first matrix represents the number of trips by forklifts between the machines.

$$
\begin{array}{c}
\begin{array}{cccc} W & X & Y & Z \end{array} \\
\begin{array}{c} 1 \\ 2 \\ 3 \\ 4 \\ 5 \end{array}
\begin{bmatrix}
3 & 3 & 3 & 5 \\
2 & 4 & 8 & 10 \\
3 & 7 & 7 & 5 \\
8 & 8 & 6 & 2 \\
10 & 10 & 4 & 2
\end{bmatrix}
\end{array}
$$

- The second matrix represents the distance between the candidate locations and the locations of the existing machines.

- We would like to combine this information so that we can minimize the material handling costs between the new and existing machines. $c_{ij} = $ (# of trips) x (distance traveled) for all the trips between the new machine locations and existing machines.
- If we multiply these two previous matrices together, we will get our overall cost assignment matrix.

$$
\begin{array}{c}
\begin{array}{cccc} W & X & Y & Z \end{array} \\
\begin{array}{c} M \\ C \\ P \\ N \end{array}
\left[\begin{array}{cccc}
190 & 230 & 226 & 228 \\
185 & 235 & 207 & 147 \\
154 & 180 & 136 & 114 \\
71 & 85 & 65 & 47
\end{array} \right]
\end{array}
$$

This is the cost matrix assignment we would like to solve with AI2.

2.14.3 Mathematical Model

The general mathematical model is called the linear assignment problem.

$$\text{Minimize } Z = \sum_{i=1}^{n} \sum_{j=1}^{n} c_{ij} x_{ij} \tag{2.12}$$

$$\text{(each machine assigned) } \sum_{j=1}^{n} x_{ij} = 1 \ \forall \ i \text{ machines} \tag{2.13}$$

$$\text{(one machine in each location) } \sum_{i=1}^{n} x_{ij} = 1 \ \forall \ j \text{ locations} \tag{2.14}$$

$$x_{ij} = \{0, 1\} \tag{2.15}$$

- The above is a linear integer programming problem which is a special case of the transportation model where the number of sources equals the number of destinations $m = n$ and each source has $s_i = 1$ and each demand point has $d_j = 1$.
- The number of basic variables is $2m - 1$.

2.14.4 Algorithm

We shall do an exhaustive search for this problem and show how to use the list structures of AI2 to effectuate the search for the optimal solution. We shall show a version of the Hungarian approach in the next section. Since $n = 4$ for our problem, we shall examine all $n! = 24$ solutions. One of the keys for AI2 is the search of the list of solutions after we set up the tabular list of the twenty-four solutions in a table. Figure 2.47 illustrates a sample of the core list blocks needed for computing the optimal solution.

Figure 2.49 illustrates the blocks.

2.14.5 Demonstration

The algorithm works very well, is very fast, and efficient. Figure 2.48 illustrates the input and final solutions for the App as indicated. Other algorithms for solving assignment problems will be treated in the Appendix.

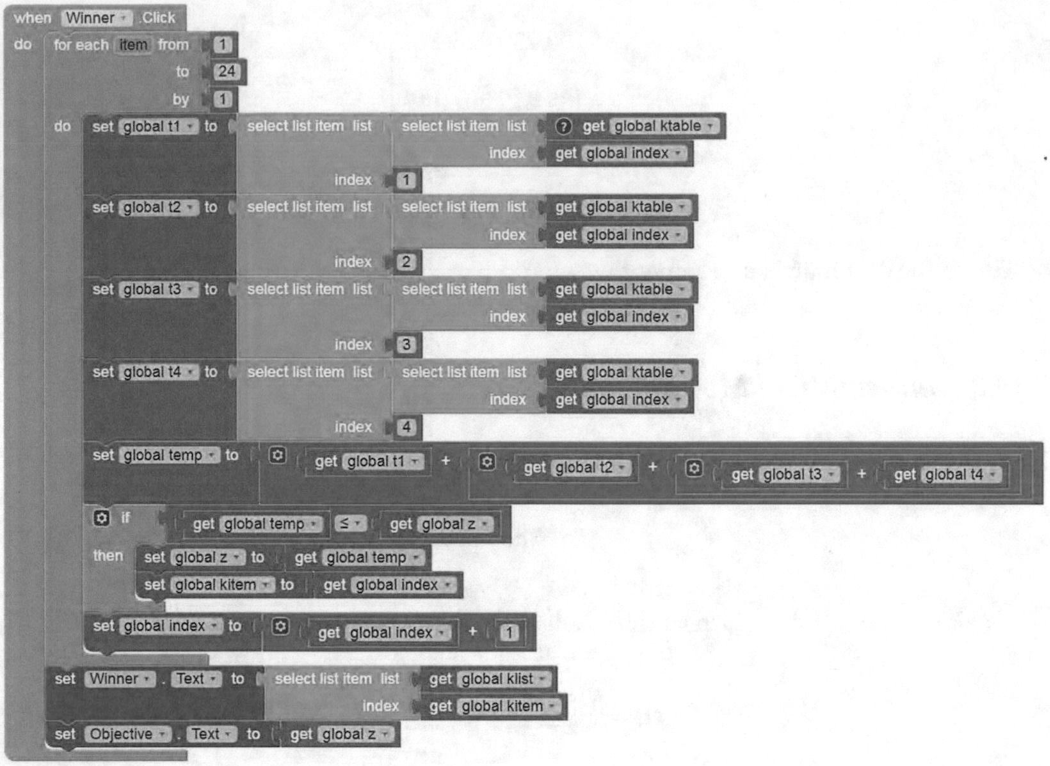

Fig. 2.47 Assignment Algorithm

2.14.6 Algorithm

A modified version of the Hungarian Assignment Algorithm is used to find the optimal assignment. This algorithm is discussed in the textbook by Taha. Here is a snapshot of the blocks used in the coding scheme. The AI2 code was developed by Lily Thomas based upon her high school experience, class of 2012. It is interesting that in most commercial applications of the Assignment problem, Linear Programming is utilized rather than the Hungarian algorithm.

2.14.7 Demonstration

As a demonstration of the approach, the App was developed for a high school track team competition rather than a swim meet but obviously the assignment problem is the same in that we have five possible candidates for four different events. Given the input data, the App correctly finds the best assignment for minimizing the total event times.

Figure 2.50 demonstrates the App for the sample problem.

So in this section of the chapter, we have given background of the Matching and Assignment problems and a basic enumeration algorithm for its solution along with a demonstration of the Hungarian algorithm which is a polynomial running time algorithm. In the Appendix, we will utilize Linear Programming to solve such problems. Now let's examine another type of Matching and Assignment problem.

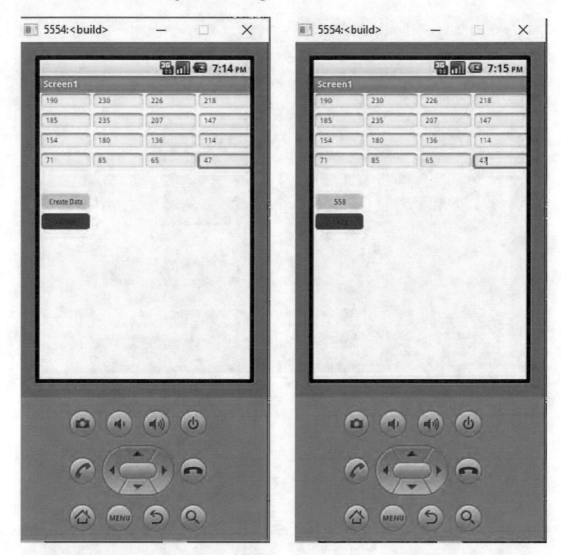

Fig. 2.48 Linear Assignment Input App and Solution

2.15 Water Polo Starting Team Assignment

This is a complex assignment/matching problem and demonstrates the problem with entering lots of input data on the phone. There are various ways in which one could pre-load some of the data so it is not so arduous. Alexander Niemeyer, Coach of UMass Water Polo, and a class member of MIE 379, programmed the App.

2.15.1 Introduction

In this App problem, there are different strengths and weaknesses associated with each potential team member. The criteria are used to develop a balanced starting line-up. The criteria/attributes are described in the panel of pictures in Figure 2.51.

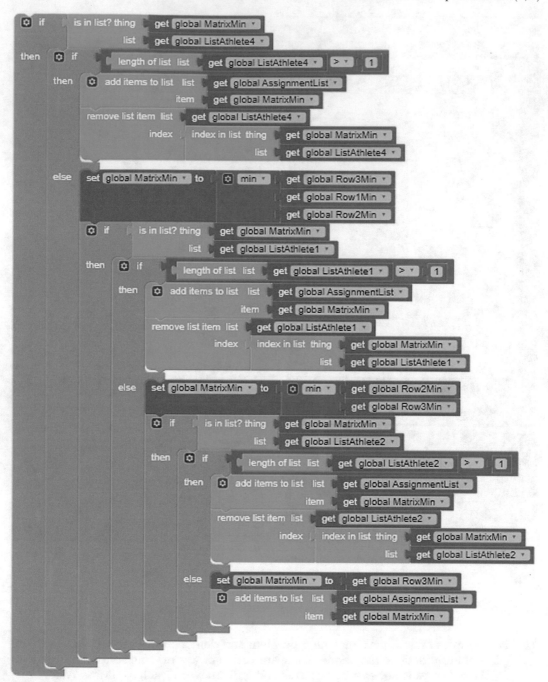

Fig. 2.49 Partial Set of Blocks for the Assignment Algorithm

2.15.2 Problem

Given the criteria, Figure 2.52 shows the numerical values for the different players and the criteria used to evaluate them. The problem posed by this App is essentially an evaluation problem that is typical in Decision Analysis.

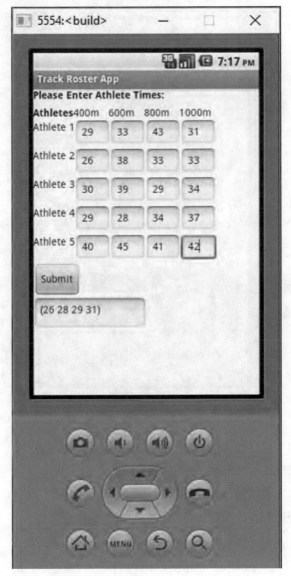

Fig. 2.50 Solution App for the Assignment Algorithm

2.15.3 Mathematical Model

Exhaustive search was used here to solve the problem. The Relative Value technique would also be an alternative solution methodology; see Exercises at the end of this chapter. Figure 2.53 illustrates the complex programming blocks.

2.15.4 Algorithm

Complete enumeration was used here to identify the optimal solution. For the data input, the App yielded the following solution:

$$Starting\ Players = \{12, 11, 2, 9, 3, 4\}$$

SPEED
Measured by a 50 yd sprint, times are normalized to a 10 point scale algorithmically:
$((S-x)/(S-F))*10$
where F is fastest time and S is slowest time.

POWER
Power is aggregated from time to empty 5 gallon jug of water while treading, 10yd sprint time, and time to complete 12 pull-up burpees.

STAMINA
Measured by time to complete test set (normalized same as speed measurement) e.g.:
50 Pushups
500m Row
50 Squat Jumps
200 m Swim
50 Burpees

SHOT BLOCKING/GENERAL DEFENSE
Measured by ranking of steals, blocks, and stops in game or best out of 10 in one on one drill.

SET
Measured by successes out of 10 attempts in scrimmage.

SHOOTING
Measured by successes out of 10 shot attempts on goalie or sniper net.

EXPERIENCE
Measured by game time or semesters playing water polo.

Fig. 2.51 Water Polo Criteria/Attributes

2.15.5 Solution App

Some selections (#12 for instance) are obvious, but other ones, such as #2 (Charles), were not obvious. The selection of #2 as a starter was in fact a selection made later in the UMass Water Polo's season, to the advantage of the team.

2.15.6 Evaluation

The program reveals interesting complexities in the varying opponents' abilities and how they relate to UMass's players' abilities. The individual opponent statistics were changed to a team aggregate, as even though a coach would want his players to stay man-on during a game, a man-on organization tends to deteriorate in a matter of minutes. Additionally, due to the lack of data on the opposing teams' players, the metrics became (for the most part) subjective evaluations, which negates the point of an objective optimization. Instead, aggregate ratings for each team were used, but ultimately provided little change.

The App Inventor program worked without a glitch but due to the volume of data involved, it was impractical to do multiple calculations on the simulator (it took too long to enter all 54 data points). In general, the App Inventor program can still be developed to make a more perfect team by analyzing more combinations and even more statistics or categories (this program, at maximum, took 144 data points and 432 iterations). However,

METHOD

	OUR TEAM									Opponent
	SPEED	POWER	STAM	D	SET	SHOT	EXP		SPEED	6
Robert	3	3	6	1	2	7	2		POWER	5
Charles	4	7	3	6	1	6	7		STAM	8
Floyd	2	7	1	6	4	5	5		D	4
Stephen	4	1	4	1	6	6	6		SET	3
Rickie	7	6	2	5	5	3	3		SHOT	5
Alonzo	5	3	2	4	1	6	3		EXP	4
Waylon	6	7	5	1	6	1	4			
Richie	7	2	6	3	6	3	3			
Rigoberto	3	6	4	7	3	2	6			
Jerome	5	2	2	4	1	3	3			
Sean	4	5	4	7	5	7	4			
Alex	9	10	10	9	9	9	10			

The data was first collected and normalized. This problem is an unique assignment problem, as in order to customize the program to the attributes of the opponent, a variable objective function was used The idea is that the program must be able to evalute the each of the six opponents and select an individual for an optimized team based on complimentary skills. E.g., a weak swimming opponent is offesively optimally matched by a strong swimmer, but the strong swimmer might be required deffensively to be matched to a strong swimmer depending on the the differential between the strong-strong pariring and the strong-weak pairing. For simplicity in AppInventor, the opposing team is assumed to be of uniform skill. Both programs then compute the efficacy of each player based on the variable or uniform opponent's function.

Fig. 2.52 Water Polo Inputs App

in some ways, water polo is more exciting because it is chaotic, so perhaps this level of investigation is enough.

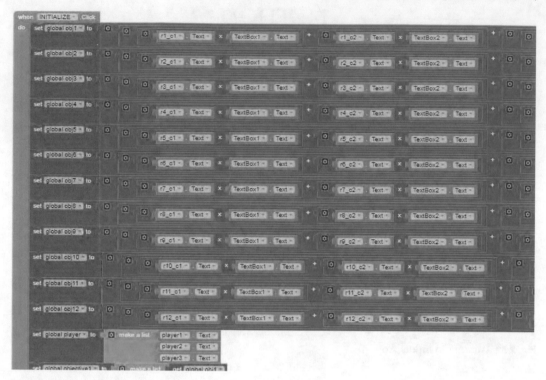

Fig. 2.53 Water Polo Blocks App

2.16 Exercises

1. **Relative Value Method:**
   Probably one of the most useful algorithms for decision-making is that of the *Relative Value* approach for choosing between alternatives in an objective manner. This first problem was developed in the first Operations Research textbook by Churchman and Ackoff [3]. It is basically a linear equation approach which solves the problem without a simplex-type algorithm but examines the underlying linear relationship between the attributes and the alternatives.
   It essentially requires a matrix data structure interrelating the attributes and alternatives.

Criteria/Attributes	Weights w	Alternatives x^1	x^2	...	x_j	...	x^r
X_1	w_1	f_1^1	f_1^2		f_1^j		f_1^r
X_2	w_2	f_2^1	f_2^2		f_2^j		f_2^r
⋮	⋮				⋮		
X_i	w_i	f_i^1	f_i^2		f_i^j		f_i^r
⋮	⋮				⋮		
X_p	w_p	f_p^1	f_p^2		f_p^j		f_p^r
Totals	$\sum w_i = 1$	V^1	V^2	...	V^j	...	V^r

Let's demonstrate an example.

Criteria/Attributes	Weights w	x^1 = New Facility	x^2 = Renovation	x^3 = Do nothing
X_1 = Energy Systems	$w_1 = 9$	9	7	1
X_2 = Office Space	$w_2 = 8$	9	7	1
X_3 = Flexibility	$w_3 = 7$	8	7	1
X_4 = Site Expansion	$w_4 = 6$	8	7	0
X_5 = Accessibility	$w_5 = 5$	7	6	0
X_6 = Image	$w_6 = 5$	8	5	1
X_7 = Acquisition \$	$w_7 = 7$	1	6	9
X_8 = Construction \$	$w_8 = 7$	6	4	9
X_9 = Operating \$	$w_9 = 8$	6	3	3
Sub-Totals	$\sum w_i = 62$	394	392	179
Grand Totals	$\frac{\sum w_i x_i^j}{\sum w_i}$	6.35	6.32	2.88

So we see that the first alternative is best.

Type I   In mathematical symbols, the formula is

$$V^j = \frac{\sum_{i=1}^p w_i x_i^j}{\sum_{i=1}^p w_i} \tag{2.16}$$

Develop an App that employs the matrix above and Type I scoring function so that you could use your App to evaluate a set of alternatives against a set of attributes/criteria like the one just presented.

2. **AHP example:**
Utilizing the Analytical Hierarchy Process (AHP), develop an App for determining an optimal diet for your living situation. Identify the attributes/criteria relevant for your situation along with relevant alternatives. Please see the solution process in the earlier parts of this chapter for guidance.

3. **TSP Tour Creation:**
   Create a Traveling Salesperson (TSP) Tour of one of your favorite cities to visit, including the important sites, monuments, museums, bodies of water, etc. based upon a user's preferences for visiting the sites along with the times to visit the sites, including travel time. You should factor in an overall time budget to restrict the length of the tour.

4. **Shortest Path Problem:**
   Develop a Shortest Path App for an application problem you are familiar with. Besides the approach in this chapter, there is another App based upon a distance matrix as input described in the Appendix to this book which is solved with the NEOS server.

5. **Restaurant Selection Process:**
   Take the Restaurant Selection App as a template and carry it out for selecting restaurants and gather data about prices and the menus based upon the type of cuisine available in the area around your campus or living situation.

6. **Minimum Spanning Tree**

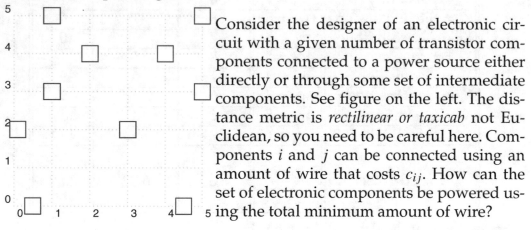

Consider the designer of an electronic circuit with a given number of transistor components connected to a power source either directly or through some set of intermediate components. See figure on the left. The distance metric is *rectilinear or taxicab* not Euclidean, so you need to be careful here. Components i and j can be connected using an amount of wire that costs c_{ij}. How can the set of electronic components be powered using the total minimum amount of wire?

3

Linear Programming $\sum c_j x_j, x_j \geq 0 \; \forall j$

Overview One of the most fundamental approaches to optimization problems is that of Linear Programming. While it is very general and has wide applicability, a number of important theoretical properties are part of LP which carry over to other optimization problems, *viz.* development of upper and lower bounds, linear approximations, simplex type approaches, etc.

An example appears in Figure 3.1. Minimize($Z = -4*x - 5*y, 0 \leq x, 0 \leq y, x + 2*y \leq 6,$ $5*x + 4*y \leq 20$); *Minimize* $Z = -19., x_1 = 2.667, y = 1.667$. Solution is green dot in the middle of plane.

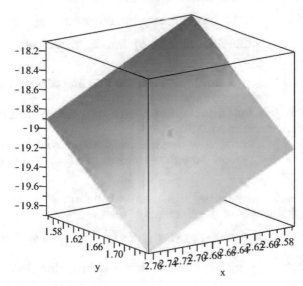

Fig. 3.1 Linear Program

Keywords: Mathematical Models, Linear Programming, Convexity

© Springer Nature Switzerland AG 2021
J. MacGregor Smith, *Combinatorial, Linear, Integer and Nonlinear Optimization Apps*,
Springer Optimization and Its Applications 175,
https://doi.org/10.1007/978-3-030-75801-1_3

True optimization is the revolutionary contribution of modern research to decision processes.

—GEORGE DANTZIG

If you optimize everything, you will always be unhappy.

—DONALD KNUTH

The 10×10 assignment problem is a linear program with 100 nonnegative variables and 20 equation constraints (of which only 19 are needed). In 1953, there was no machine in the world that had been programmed to solve a linear program this large!

—HAROLD KUHN

3.1 Introduction

In general, Linear Programming (LP) has the following characteristics.

- An objective function $f(x_1, \ldots, x_n)$ of $(x_1, \ldots, x_n)$ is a **linear function** if and only if for some set of parameter constants $c_1, c_2, \ldots, c_n, f(x_1, \ldots, x_n) = c_1 x_1 + c_2 x_2 + \ldots + c_n x_n$. Some examples of linear and nonlinear objective functions are as follows:

  - $f(x_1, x_2) = 3x_1 + 2x_2$ is a linear function but $f(x_1, x_2) = 3x_1^2 + 2x_1 x_2^2$ is not a linear function of x_1 and x_2.

- For any linear objective function $f(x_1, \ldots, x_n)$ and any right hand side number b, the inequalities which will constrain the solutions for our objective $f(x_1, \ldots, x_n) \leq b$ and $f(x_1, \ldots, x_n) \geq b$ are **linear inequalities.** Some examples are

  - $2x_1 + 3x_2 \leq 4$ and $4x_1 - 3x2 \geq 1$ are linear inequalities but $x_1^3 x_2 - 2x^2 \geq 3$ is not.

- A Linear Programming problem (LP) is an optimization problem for which we do the following:

  1. We attempt to maximize (or minimize) a linear function of the decision variable. This function is called the objective function.
  2. The values of the decision variables must satisfy a set of constraints. Each constraint is either an inequality or an equality.
  3. A sign restriction is associated with each decision variable. For any x_i, the sign restriction specifies that x_i must be either nonnegative ($x_i \geq 0$) or unrestricted in sign (*urs*).

So if we assume these linear relationships, then we can formulate the following LP model. We need some basic definitions of the elements of the LP. These include

- *Decision Variables:* Usually denoted as vector of variables with subscripts although they need not be subscripted $\mathbf{x} = (x_1, \ldots, x_n)$
- *Objective Function:* A performance variable that is either minimized or maximized and usually denoted as $f(\mathbf{x})$
- *Constraints:* A qualitative objective that is usually a bounding mechanism on the decision variables, usually denoted as $g(\mathbf{x}) \leq 0; h(\mathbf{x}) = 0$

Once we have these component elements, then we can formulate algebraically a general mathematical programming problem model as

$$\text{Max or Min } f(\mathbf{x}) \tag{3.1}$$
$$subject\ to : \text{inequalities } g_i(\mathbf{x}) \leq 0 \tag{3.2}$$
$$\text{equalities } h_i(\mathbf{x}) = 0 \tag{3.3}$$
$$\text{nonnegativity } \mathbf{x} \geq 0 \tag{3.4}$$

Some of the key assumptions one must make in setting up LPs are the following:

- **Proportionality:** the objective functions and constraints are linear in the given parameters.
- **Additivity (Separability):** the decision variables are separable and independent.
- **Divisibility:** the decision variables are continuous.
- **Certainty:** the values of the objective function and constraints and parameters are know with certainty.

Given these assumptions and the general linear programming structure, let's examine some LP Apps. Normally, these linear programs are solved with what is called the simplex algorithm. The Apps described in the following pages will not always use the simplex algorithm for their solution but rely on the linear properties for their solution.

3.2 Diabetes Problem

Brendan Frankfort, a student in the 2016 class, developed an App for measuring the response to his Type I Diabetes. Persons with Type I Diabetes must take insulin to cope with their illness.

3.2.1 Introduction

Brendan is one of about 1.25 million people in the world that have Type 1 diabetes. Diabetes is difficult to cope with manually, which is why medical devices such as insulin pumps are so effective. An insulin pump is a medical device that monitors and delivers insulin automatically to keep the blood sugar level constant during exercise, eating, sleeping, etc. Unfortunately this device is very expensive, even if the user is lucky enough to have an insurance company that helps pay for it. This App is designed for diabetics that do not have an insulin pump.

3.2.2 Problem

The App would track the users' level of physical activity (light, moderate, or hard) and the length of time of the activity. It would also track food intake and how quickly the food goes into the bloodstream. For example, high sugar candy takes a lot less time to bring the blood sugar up than pasta or bread. This App would need inputs of initial statistics from the user such as target blood sugar range, high and low limits to adjust blood sugar, insulin to carb ratio, insulin to blood sugar ratio, correction factor, etc. Since these statistics take a while to develop for a new diabetic, this App is better for experienced diabetics. After the user inputs their initial statistics, the App would be ready for predicting the demand that the user encounters. The user inputs food, exercise, blood sugar stats in the App and it outputs the amount of insulin or food that the user would need to consume to regulate his/her blood sugar.

In Figure 3.2, we see the inputs to the App.

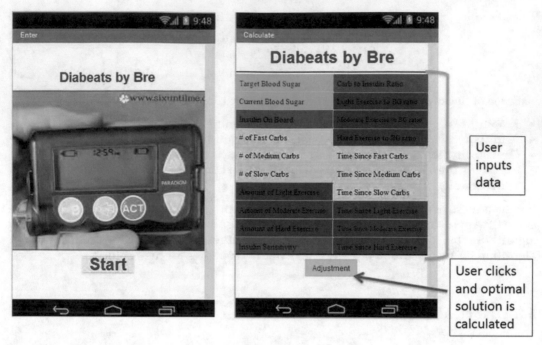

Fig. 3.2 Diabetes App Inputs

3.2.3 Mathematical Model

Linear Programming was used to model the problem, since the units of the decision variables are continuous. There are also many constraints. Figure 3.3 illustrates part of the Mathematical Model.

3.2.4 Algorithm

Because of the size of the LP, the NEOS server was utilized to solve the problem.

3.2.5 Solution App

In Figure 3.4, we illustrate the programming blocks for the Diabetes App.

3.2.6 Evaluation

The output of the App is the amount (mg/dL) of blood sugar that needs to be adjusted. However, people do not directly adjust that number. Once the blood sugar level is calculated, the user takes the information and makes a decision whether the adjustment is positive or negative. If the solution is positive, the user would divide that number by the insulin sensitivity to get the units of insulin that the user should receive. However, if the solution is negative, the user would multiply the number by Insulin Sensitivity and divide by Carb to

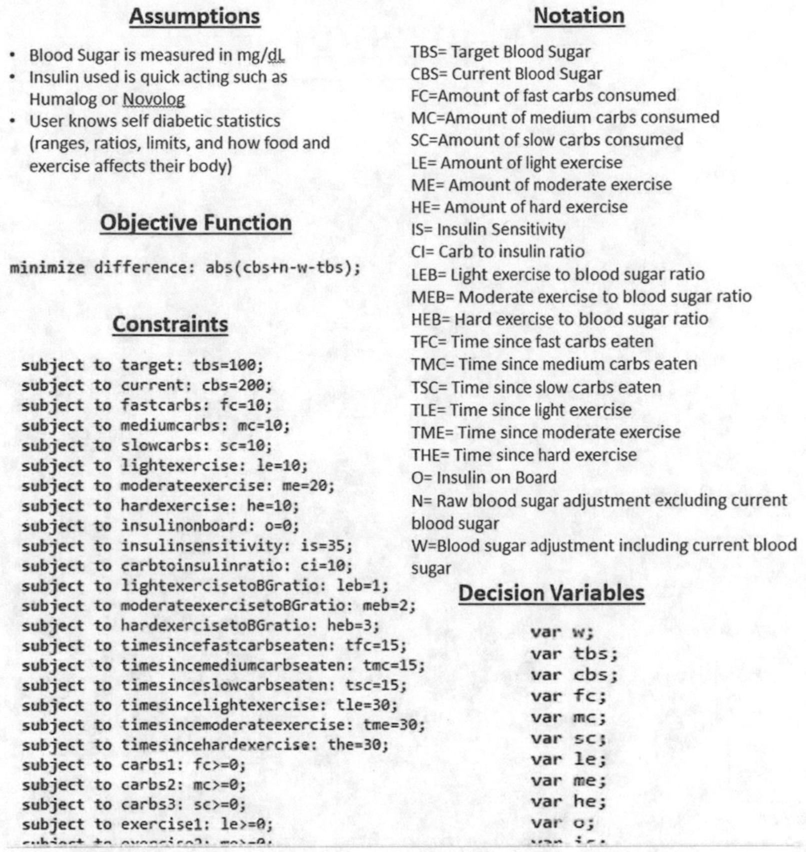

Assumptions

- Blood Sugar is measured in mg/dL
- Insulin used is quick acting such as Humalog or Novolog
- User knows self diabetic statistics (ranges, ratios, limits, and how food and exercise affects their body)

Objective Function

```
minimize difference: abs(cbs+n-w-tbs);
```

Constraints

```
subject to target: tbs=100;
subject to current: cbs=200;
subject to fastcarbs: fc=10;
subject to mediumcarbs: mc=10;
subject to slowcarbs: sc=10;
subject to lightexercise: le=10;
subject to moderateexercise: me=20;
subject to hardexercise: he=10;
subject to insulinonboard: o=0;
subject to insulinsensitivity: is=35;
subject to carbtoinsulinratio: ci=10;
subject to lightexercisetoBGratio: leb=1;
subject to moderateexercisetoBGratio: meb=2;
subject to hardexercisetoBGratio: heb=3;
subject to timesincefastcarbseaten: tfc=15;
subject to timesincemediumcarbseaten: tmc=15;
subject to timesinceslowcarbseaten: tsc=15;
subject to timesincelightexercise: tle=30;
subject to timesincemoderateexercise: tme=30;
subject to timesincehardexercise: the=30;
subject to carbs1: fc>=0;
subject to carbs2: mc>=0;
subject to carbs3: sc>=0;
subject to exercise1: le>=0;
subject to exercise2: me>=0;
```

Notation

TBS= Target Blood Sugar
CBS= Current Blood Sugar
FC=Amount of fast carbs consumed
MC=Amount of medium carbs consumed
SC=Amount of slow carbs consumed
LE= Amount of light exercise
ME= Amount of moderate exercise
HE= Amount of hard exercise
IS= Insulin Sensitivity
CI= Carb to insulin ratio
LEB= Light exercise to blood sugar ratio
MEB= Moderate exercise to blood sugar ratio
HEB= Hard exercise to blood sugar ratio
TFC= Time since fast carbs eaten
TMC= Time since medium carbs eaten
TSC= Time since slow carbs eaten
TLE= Time since light exercise
TME= Time since moderate exercise
THE= Time since hard exercise
O= Insulin on Board
N= Raw blood sugar adjustment excluding current blood sugar
W=Blood sugar adjustment including current blood sugar

Decision Variables

```
var w;
var tbs;
var cbs;
var fc;
var mc;
var sc;
var le;
var me;
var he;
var o;
var is;
```

Fig. 3.3 Diabetes Mathematical Model

Insulin ratio to get the amount of carbs that the user should consume. This App would help improve blood sugar levels and therefore would improve A1C levels, which is a test that shows the user how effectively they control their diabetes. In Figure 3.5 we see the outputs for the App.

3.3 Equalization of Runout Times (ERT) Oil Delivery App

A closely related problem to the Stowe Cycle TSP routing problem which we reviewed earlier is an inventory resource allocation problem where the allocation of the resource to the demand points $V = \{v_1, v_2, \ldots, v_n\}$ is such that it maximizes the time at which the group of demand points will next be scheduled. This is actually classified as a capacitated inventory problem on a graph which turns out to be a linear programming problem.

3.3.1 Introduction

Suppose we have a heating oil delivery problem and we wish to allocate the amount of oil to each of $n-$households with demand point x_i so that the allocation of oil will ensure the maximum time to re-supply, *i.e.* all households in a given area will run out of oil at the same time.

App Inventor

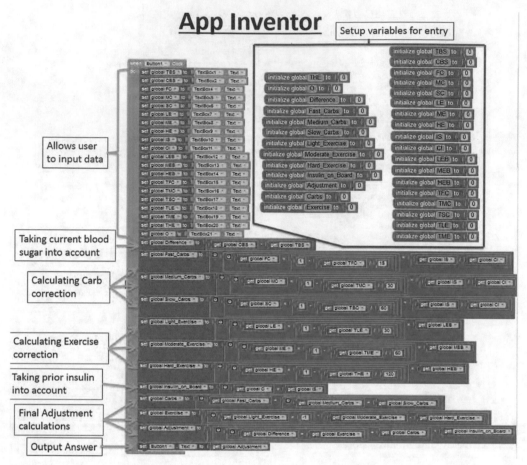

Fig. 3.4 Diabetes App with Blocks

AMPL Output Example

```
ampl: model Diabeats.mod;
ampl: solve;
Solution determined by presolve;
objective difference = 0.
ampl: display cbs, n, w, tbs;
cbs = 200
n = 1.25
w = 101.25
tbs = 100
```

Fig. 3.5 Diabetes Output

If we don't have the households running out of oil at the same time, the truck must be dispatched to the area multiple times, whereas if they run out at the same time, then we can dispatch the truck once. What will be the allocation of oil to each household so that you maximize the time between deliveries? Figure 3.6 illustrates the problem.

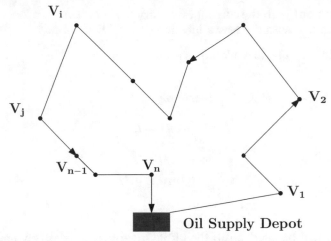

Fig. 3.6 ERT Oil Supply Problem

3.3.2 Problem

Fundamentally, this is a Linear Programming (LP) problem and is one of the basic set of topics in the Taha textbook which we will study, but actually we can solve it more readily without the simplex method. Below are some of the variables and parameters we need to build a mathematical model of the problem.

3.3.3 Mathematical Model

The notation needed for the ERT problem is described below.

Variable Description	
$D_i :=$	Demand at household i
$E_i :=$	Initial inventory supply at the household
$K :=$	Capacity of Delivery Vehicle
$n :=$	n-households
$t :=$	Time to runout of the inventory item and for each household ($t_i = x_i/D_i$)
$x_i :=$	Decision variable for the allocation at a household

Our Linear Programming (LP) formulation appears in the following statements where we maximize the runout time subject to the demand and current inventory levels and the capacitated supply constraints of the perishable item. In the formulation below, the initial inventory level is 0.

$$Maximize\, t \tag{3.5}$$

$$s.t. \sum_{i}^{n} x_{i=1} = K \tag{3.6}$$

$$\frac{(x_i + E_i)}{D_j} - t \ge 0 \tag{3.7}$$

$$x_i \ge 0\, \forall i \tag{3.8}$$

If in addition, we allow for an initial inventory, then we have the following extension:

- $w_i :=$ time to run out with the current inventory ($w_i = E_i/D_i$)
- $u_i :=$ time to run out with the new allocation ($u_i = x_i/D_i$).

Then a new problem can be formulated as an LP:

$$\text{Maximize} \, t \tag{3.9}$$

$$s.t. \ \sum_i^n u_i D_i = K \tag{3.10}$$

$$w_i + u_i - t \geq 0 \tag{3.11}$$

$$u_i, w_i, t \geq 0 \ \forall j \tag{3.12}$$

In general, LP must be used, but if the initial inventory is relatively small, the ERT rule works:

$$w_i + u_i = t \rightarrow u_i = t_i - w_i, \text{and} \ \sum u_i D_i = \sum (t - w_i) D_i = t \sum D_i - \sum w_i D_i = K \tag{3.13}$$

and

$$t = (H + \sum w_i D_i)/(\sum D_i) = (K + \sum E_i)/(\sum D_i) = t \tag{3.14}$$

and

$$u_i = t - w_i \rightarrow (K + \sum E_i)/(\sum D_i) - E_i/D_i \tag{3.15}$$

then the optimal allocation for a household is

$$x_i = D_i t - E_i \tag{3.16}$$

and this allocation process will work as long as $x_i \geq 0$.

The App can be directly programmed as described in the next section of the book, since the solution methodology is relatively straightforward and depends upon the data about the demands and inventory and the formula above.

3.3.4 Algorithm

We do not have to use the simplex algorithm of Linear Programming per se to solve the problem because the structure of the problem can be utilized for a straightforward solution. Figure 3.7 illustrates a sample of the block programming of the Equalization of Run Out Times for Oil (ERTGO) App. The blocks detail the summing of the demands. Because of the size of the designer screen, seven households can be the maximum used in the problem. A different screen design could obviously handle more demand points. The remaining blocks are available with the App on the website.

3.3.5 Demonstration

Figure 3.8 illustrates the final implementation of the ERTGO App along with a demonstration for the following data and parameters. For example, let's say that we have $N = 4$ and the demand vector for the households is $D = \{120, 180, 900, 50\}$; the current inventory supply is $E = \{30, 120, 300, 20\}$; we have $K = 1000$ gallons capacity for the truck, Running the App, we find that $t = 1.1760$ time periods and the allocation to the households is

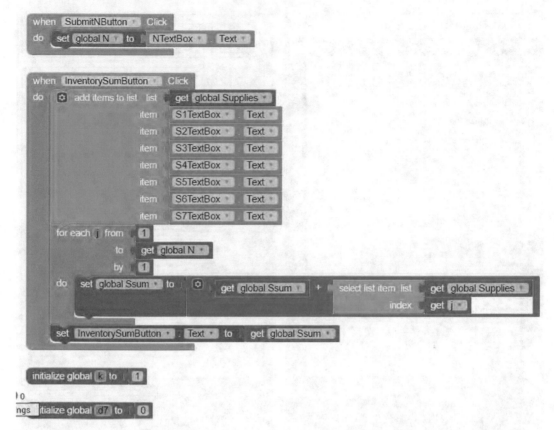

Fig. 3.7 ERT Blocks

$X = \{111.12, 91.68, 758.4, 38.8\}$. The algorithm takes the capacity of the truck and fills the household demands very compactly.

3.3.6 Evaluation

The App is fairly easy to use and it is very effective. For another application where LP is indirectly used, we discuss a location problem in the next section of the book which can be transformed into a linear program.

3.4 Pinball Weber Machine Location App

This is a location problem with a rectilinear (taxicab) metric which can be formulated as a Linear Program. The solution will not explicitly use LP per se, but will use the separability properties and nature of the problem to generate a linear methodology for a solution. A Euclidean form of this problem will be dealt with in Chapter 5 since the Euclidean version is essentially a nonlinear programming problem.

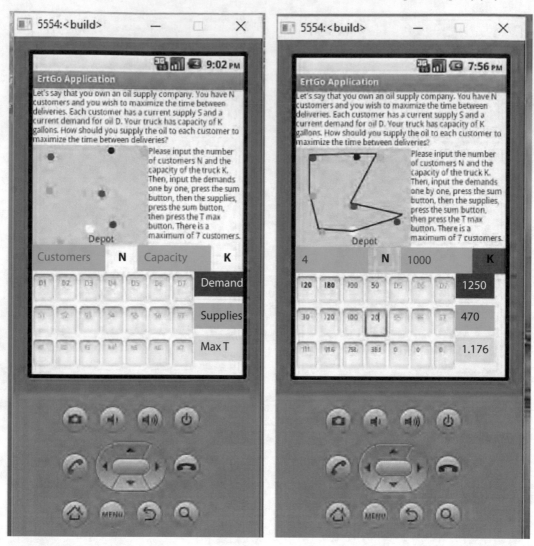

Fig. 3.8 ERTGO App Success

3.4.1 Introduction

The Steiner/Weber problem is a famous location problem that goes back to the 18th century. In fact, I worked on this problem for my Ph.D. dissertation. Most location problems are nonlinear in nature since they require a Euclidean distance function. The problem is to locate a new facility in relation to some existing facilities where we wish to minimize the total distance traveled between the new facility and the existing facilities.

3.4.2 Problem

Let's say that we want to locate a new machine within a factory where we already have four existing machines located within the building at the Cartesian points $(8, 5), (4, 2), (11, 8)$, and $(13, 2)$. In the manufacturing facility, there are weights of importance representing $\mathbf{w} = \{9, 6, 4, 12\}$, the expected number of trips per week for a material handling system between the new machine and the current machines. Minimizing the total weighted traffic flow is our objective. Figure 3.9 illustrates the location problem.

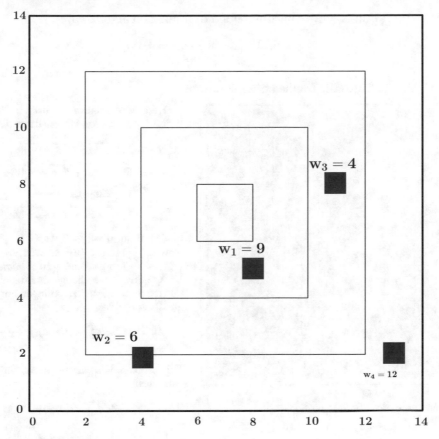

Fig. 3.9 Layout of Factory

3.4.3 Mathematical Model

We wish to find the location of the new machine so as to minimize the total distance traveled between the new machine and the existing machines. Let's first analyze the problem where we assume Euclidean distance is used to travel between the new machine $i(x, y)$ and the j−existing machines.

$$\text{Minimize Z} = \sum_j w_j dist(i, j) \tag{3.17}$$

So we will also have an unconstrained optimization problem but instead of one variable, we will have two variables (x, y). In a subsequent model, we will use rectilinear distance.

The objective function is

$$Z = 9\sqrt{x^2 - 16x + 89 + y^2 - 10y} + 6\sqrt{x^2 - 8x + 20 + y^2 - 4y} + \tag{3.18}$$

$$4\sqrt{x^2 - 22x + 185 + y^2 - 16y} + 12\sqrt{x^2 - 26x + 173 + y^2 - 4y} \tag{3.19}$$

The derivatives are very complicated and the nonlinear equations cannot be separated in the decision variables x, y so we have the following:

$$-\frac{9(2x - 16)(2y - 10)}{4(x^2 - 16x + 89 + y^2 - 10y)^{3/2}} - \frac{3(2x - 8)(2y - 4)}{2(x^2 - 8x + 20 + y^2 - 4y)^{3/2}} \tag{3.20}$$

$$-\frac{(2x - 22)(2y - 16)}{(x^2 - 22x + 185 + y^2 - 16y)^{3/2}} - \frac{3(2x - 26)(2y - 4)}{(x^2 - 26x + 173 + y^2 - 4y)^{3/2}} \tag{3.21}$$

The actual solution is (via an unconstrained optimization algorithm)

$$[Z = 116.117, [X = 8.737, Y = 4.451]]$$

Figure 3.10 illustrates the Mathematica solution.

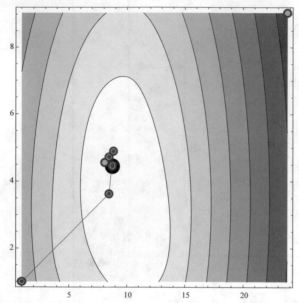

Fig. 3.10 Mathematica Steiner Solution

Here, Mathematica was used to find the solution. Optimization 'UnconstrainedProblems'

$$f = 9\sqrt{x^2 - 16x + 89 + y^2 - 10y +} \quad (3.22)$$
$$6\sqrt{x^2 - 8x + 20 + y^2 - 4y +} \quad (3.23)$$
$$4\sqrt{x^2 - 22x + 185 + y^2 - 16y +} \quad (3.24)$$
$$12\sqrt{x^2 - 26x + 173 + y^2 - 4y} \quad (3.25)$$

FindMinimumPlot$[f, \{\{x, 1\}, \{y, 1\}\},$ Method $\rightarrow$ Newton$]$ $\{\{116.117, \{x \rightarrow 8.73765, y \rightarrow 4.4518\}\},$ $\{"Steps" \rightarrow 7, "Function" \rightarrow 10, "Gradient" \rightarrow 10\}$

Actually, if we change our distance function from Euclidean to rectilinear, our problem greatly simplifies.

$$\text{Minimize } Z = \sum_j w_j(|x - a_{j1}| + |y - a_{j2}|)$$
$$(3.26)$$

This problem is separable in x, y the decision variables:

$$Z_x = \sum_j w_j(|x - a_{j1}|) \quad (3.27)$$
$$Z_y = \sum_j w_j(|y - a_{j2}|) \quad (3.28)$$

We actually can show that these two separate functions are convex and the optimal (x, y) location is the median sum of the weights of the existing sorted facilities on the x and y coordinates. Let's take our example problem and examine this convexity issue:

	Original Coordinates			sorted x_1 dimen.		sorted x_2 dimen.	
j	a_{j1}	a_{j2}	w_j	a_{j1}	w_j	a_{j2}	w_j
1	8	5	9	4	6	2	6
2	4	2	6	8	9	2	12
3	11	8	4	11	4	5	9
4	13	2	12	13	12	8	4

$$W_1(x_1) = 6|x_1 - 4| + 9|x_1 - 8| + 4|x_1 - 11| + 12|x_1 - 13| \quad (3.29)$$
$$W_1(0) = 296, W_1(4) = 145, W_1(8) = 96, W_1(11) = 93, W_1(13) = 107 \quad (3.30)$$
$$W_2(x_2) = 6|x_2 - 2| + 12|x_2 - 2| + 9|x_2 - 5| + 4|x_2 - 8| \quad (3.31)$$
$$W_2(0) = 113, W_2(2) = 51, W_2(5) = 66, W_2(8) = 135 \quad (3.32)$$

3.4.4 Algorithm

The algorithm solves the problem for the X-coordinate then the Y-coordinate because the problem is separable in the decision variables. It is very straightforward to find the solution.

- [Step 1.0:] First find the sum of the weights $\sum_j^n w_j$
- [Step 2.0:] x^* is the first a_j (sorted x-order) at which the cumulative weights reach or exceed the median value $\sum_j^n w_j/2$
- [Step 3.0:] y^* is the first a_j (sorted y-order) at which the cumulative weights reach or exceed the median value $\sum_j^n w_j/2$

In our example problem, the sum of the weights is $W = 31$ (*i.e.* median = 15.5). We sort on the x-axis first, identify the

- $x = \{4, 8, 11, 13\}$, and $w_j = \{6, 9, 4, 12\}$ then $x^* = 11$
- while $y = \{2, 2, 5, 8\}$ and $w_j = \{12, 6, 9, 4\}$ then $y^* = 2$,
- the optimal location is $(\mathbf{x}, \mathbf{y}) = (\mathbf{11}, \mathbf{2})$ with $Z = 144$.

3.4.5 Demonstration

Figure 3.11 is the final realization of the App. The red ball represents the optimal location for the given problem, while the black dots represent the existing facilities. The green dot moves around the screen to show changes to the screen after the input arrangement of black dots is reset. Once the green dot hits a boundary, the new optimal location appears as the red dot.

Below in Figure 3.12 is the App and its blocks programming with AI2. The App is actually programmed to solve randomly generated problems and to show the graphic capabilities of the AI2 language and phone properties. This background map-making ability in AI2 can be very useful for location optimization problems.

3.4.6 Evaluation

Some of the features of the App are

[•] It iteratively generates a new four-facility location problem with random Cartesian coordinates and weights. It also recomputes the optimal location as the red dot on the screen. One of the nice features of App Inventor is the dynamics possible with the screen.
[•] ReStart button once pushed ⇒ randomly generates a new problem.
[•] Once green ball hits boundary, a new solution is found.
[•] The App works very fast.

The App could be programmed to handle a larger number of given machines or facilities, and the choice of four is simply illustrative.

3.5 Breakfast Optimization Problem

Tim Klocker of the MIE 379 class of 2017 programmed this App. It is a straightforward use of LP. While the formulation is limited to breakfast, including other meals would follow a similar methodology.

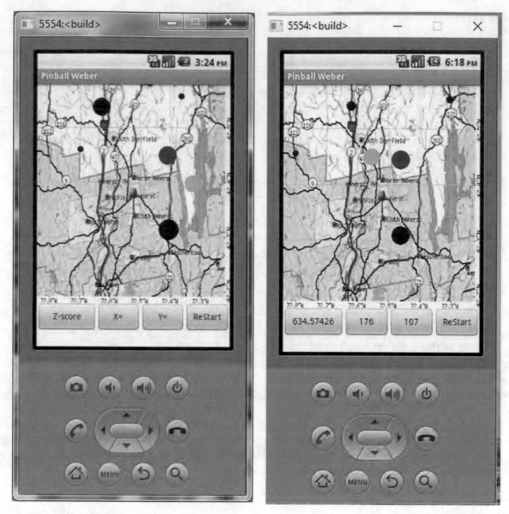

Fig. 3.11 Pinball App Input and Output Screens

3.5.1 Introduction

A study in the American Journal of Clinical Nutrition observed 12,000 individuals over 5 years on the effect of breakfast on health. The study concluded that those participants who ate breakfast had an overall decreased energy density, a lower BMI, and a higher diet quality. In the study Irregular Breakfast Eating Associated Health Behaviors: A Pilot Study Among College Students, 1,257 college students answered a questionnaire regarding their eating habits and health. The results of the study discovered only 23.8% of the sample ate breakfast every day. These statistics show that breakfast is an important meal for overall health which many college students are lacking. The purpose of the App is to determine the

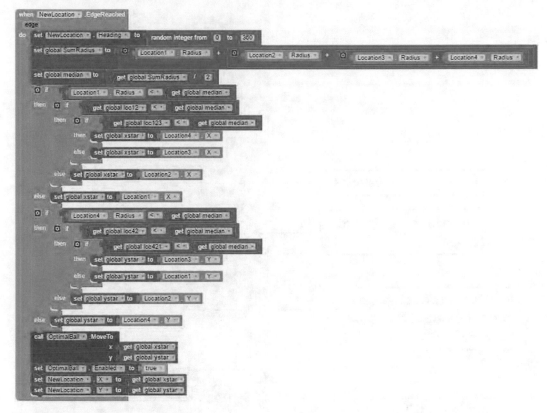

Fig. 3.12 Sample Blocks and Pinball App

optimal breakfast for a given amount of time. While the problem is a little offbeat, the use of the AI2 technology is interesting. [1] [2] [3]

3.5.2 Problem

Figure 3.13 illustrates the welcome screen.

3.5.3 Mathematical Model

Conversion Calculations Within App:

- Breakfast Calories= Daily Calories*0.35
- Protein grams to calories= grams*4

[1] Kant, Ashima K., Mark B. Andon, Theodore J. Angelopoulos, James M. Rippe. "Association of Breakfast Energy Density with Diet Quality and Body Mass Index in American Adults: National Health and Nutrition Examination Surveys, 1999–2004." American Journal of Clinical Nutrition 88, no. 5 (2008): 1396-1404.

[2] Thiagarajah, Krisha, and Mohammad R. Torabi. "Irregular Breakfast Eating Associated Health Behaviors: A Pilot Study Among College Students." Institute of Education Sciences, files.eric.ed.gov/fulltext/EJ865575.pdf

[3] Hanson, Mary. "Balancing Carbs, Protein, and Fat." Balancing Carbs, Protein, and Fat, Kaiser Permanente, 1 Mar. 2014,wa.kaiserpermanente.org/healthAndWellness/

Design Screen

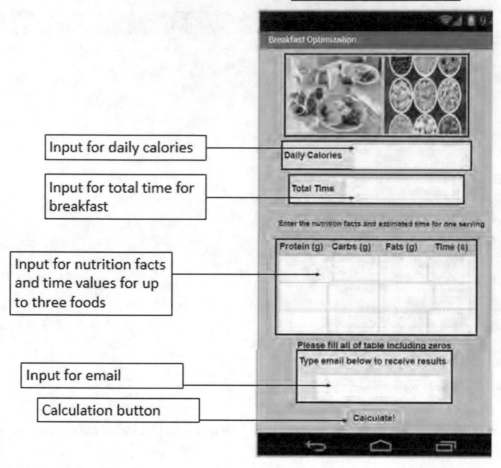

Fig. 3.13 Input for Breakfast App

- Carbs grams to calories= grams*4
- Fat grams to calories=grams*9

Decision variables: (# of servings) Breakfast Food type 1(f1), Breakfast Food type 2(f2), Breakfast Food type 3(f3).

Objective function:

$$Maximize \; S = (Servings)f1 + (Servings)f2 + (Servings)f3 \tag{3.33}$$

$$s.t. \; (protein\,cal.)f1 + (protein\,cal.)f2 + (protein\,cal.)f3 \leq \text{Breakfast Calories} * 0.2 \tag{3.34}$$

$$(protein\,cal.)f1 + (protein\,cal.)f2 + (protein\,cal.)f3 \geq \text{Breakfast Calories} * 0.12 \tag{3.35}$$

$$(carbs\,cal.)f1 + (carbs\,cal.)f2 + (carbs\,cal.)f3 \leq \text{Breakfast Calories} * 0.6 \tag{3.36}$$

$$(carbs\,cal.)f1 + (carbs\,cal.)f2 + (carbs\,cal.)f3 \geq \text{Breakfast Calories} * 0.5 \tag{3.37}$$

$$(fat\,cal.)f1 + (fat\,cal.)f2 + (fat\,cal.)f3 \leq \text{Breakfast Calories} * 0.3 \tag{3.38}$$

$$(Time)\,f1 + (Time)f2 + (Time)f3 \leq \text{Total Allotted Time.} \tag{3.39}$$

3.5.4 Algorithm

We will utilize the CPLEX Linear Program software available on the NEOS server to solve this problem. Figure 3.14 illustrates a sample of the blocks used in the program.

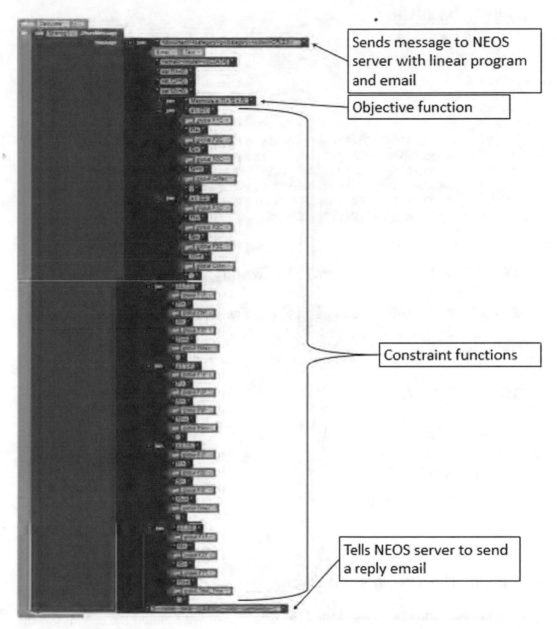

Fig. 3.14 Sample Blocks

3.5.5 Solution App

Figure 3.15 illustrates one solution to the problem for the given set of data.

AMPL Example .

```
#Decision Variables
var f1>=0;
var f2>=0;
var f3>=0;
#Objective Function
maximize s: f1+f2+f3;
#Constraints
s.t. C1: 3.6*f1+45*f2+9*f3<=210;#Maximimum Fat Constraint
s.t. C2: 5.2*f1+24*f2+96*f3<=140;#Maximum Protein Constraint
s.t. C3: 5.2*f1+24*f2+96*f3>=84;#Minimum Protein Constraint
s.t. C4: 108*f1+2.4*f2+12*f3<=420;#Maximum Carb Constraint
s.t. C5: 108*f1+2.4*f2+12*f3>=350;#Minimum Carb Constraint
s.t. C6: 30*f1+600*f2+120*f3<=1800;#Time Constraint
```

```
CPLEX 12.7.0.0: optimal solution; objective 7.0406642

3 dual simplex iterations (1 in phase I)

f1 = 3.76448

f2 = 2.69567

f3 = 0.580505

s = 7.04066
```

Fig. 3.15 Breakfast AMPL Solution

3.5.6 Evaluation (Benefits and Costs)

This App is for daily use to optimizing the user's breakfast limiting the servings to the allotted time and consuming a healthy balance of nutrients. The App could still be improved by adding a suggestion option for the user based on their preferential tendencies.

3.6 National Guard Training

This App was developed by Eric Wright, a student of MIE 379 in the class of 2015. Every year each soldier in the US Army, US Army National Guard, and US Army Reserve must conduct what is called a rifle qualification. The rifle qualification is a test to measure the proficiency of each soldier's marksmanship skills.

3.6.1 Introduction

In most combat units, these qualifications are run at the company level, meaning around 120 soldiers must take the test during the same qualification event. The event consists of 40 pop-up targets at varying distances between 50 and 300 meters. These targets are engaged from various stances and the number of targets hit indicates a soldier's score. There are three levels of proficiency: marksmen, sharpshooter, and expert. Each soldier must qualify as at least a marksmen, however, the soldier can take the test as many times as needed in the allotted training day to complete this task or improve their score. For one iteration of the test, each soldier is given 40 rounds of $NATO5.56mm$ ball ammunition; one for each target. This can put commanders in a predicament. If they order too little, some soldiers may not qualify or get the score they want. If they order too much, some of the unit's budget will be wasted on extra ammunition. Unused ammunition cannot be returned and must be spent. This can be highly wasteful and time-consuming if the training day is short.

Figure 3.16 illustrates the input screens for the App.

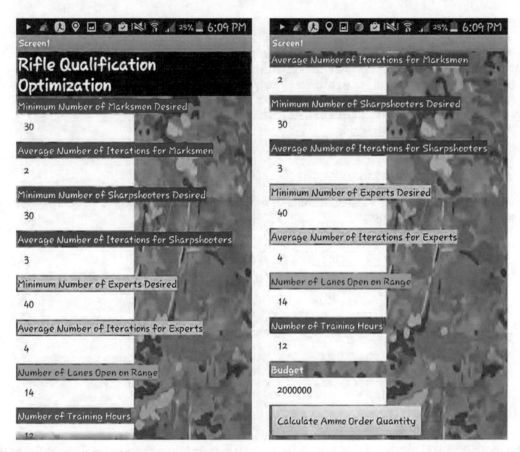

Fig. 3.16 National Guard Inputs

3.6.2 Problem

Typically one or two ranges are utilized for qualification and each range is made up of about fifteen lanes (1 lane per soldier at a time) but due to the nature of the electronic targets, lanes are often down for repairs. This App allows leaders of any sized unit to input several variables and receive an optimal quantity of ammunition to order. These parameters consist of

- *Budget*
- *Number of Training Hours*
- *Number of Lanes Open on the Range*
- *Minimum Desired Marksmen, Sharpshooters, and Experts*
- *Average Number of Iterations performed by Marksmen, Sharpshooters, and Experts*

Most of these variables will be known by the organizers of the qualification event. However, the average number of iterations can be obtained by the previous years' records as all this information is stored electronically.

3.6.3 Mathematical Model

```
Notation:
E = minimum number of experts desired
S = minimum number of sharpshooters desired
M = minimum number of marksmen desired
Ie = avg. number of iterations for experts
Is = avg. number of iterations for sharpshooters
Im = avg. number of iterations for marksmen
L = number of lanes open
T = training hours allotted
B = budget
Z = amount of rounds to order
```

   Decision Variables: $x_1 :=$ final number of marksmen $x_2 :=$ final number of sharpshooters $x_3 :=$ final number of experts
   Finally, we have the following LP:

$$\text{Minimize } Z = Im*(40)*x_1 + Im*(40)*x_2 + Im*(40)*x_3 \tag{3.40}$$

$$\text{s.t. :} Im*(.25)*x_1 + Im*(.25)*x_2 + Im*(.25)*x_3 \leq L*T \tag{3.41}$$

$$Im*(.39)*x_1 + Im*(.39)*x_2 + Im*(.39)*x_3 \leq B \tag{3.42}$$

$$x_2 \geq S \tag{3.43}$$

$$x_3 \geq E \tag{3.44}$$

$$x_j \geq 0 \,\forall j \tag{3.45}$$

3.6.4 Algorithm

The mathematical model of this App is based on a general linear programming model and as such, can be solved by a program called AMPL. When the calculate button is pushed, The App converts the input into a syntax that AMPL understands and it is emailed to the NEOS server to be solved. The answer returns as a reply to this email.

3.6.5 Solution App

Figure 3.17 illustrates the blocks programming for this App. Below is the solution from NEOS.

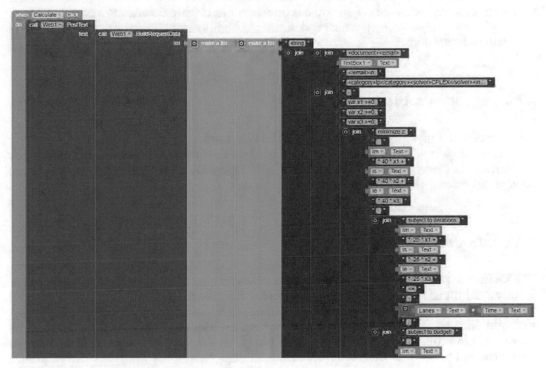

Fig. 3.17 National Guard Algorithm Blocks Programming

```
You are using the solver cplexamp.
Checking ampl.mod for cplex_options...
Checking ampl.com for cplex_options...
Executing AMPL.
processing data.
processing commands.
Executing on prod-exec-5.neos-server.org

Presolve eliminates 3 constraints.
Adjusted problem:
3 variables, all linear
2 constraints, all linear; 6 nonzeros
        2 inequality constraints
1 linear objective; 3 nonzeros.

CPLEX 12.10.0.0: threads=4
CPLEX 12.10.0.0: optimal solution; objective 13600
0 dual simplex iterations (0 in phase I)
x1 = 30
x2 = 30
x3 = 40
z = 13600
```

3.6.6 Evaluation

This optimization App is an excellent tool for leaders at the platoon, company, or even battalion level to determine how much ammunition they need to order to complete qualifications. This App allows commanders to adjust the inputs if they want to prioritize their unit's score, see if their desired outcome is time feasible, or help determine a proper budget.

Now let's examine an App which is designed to demonstrate the formulation and use of the knapsack problem.

3.7 Knapsack Problem

Another classic Linear Program problem is the Knapsack problem. It is the basis for many allocation problems. While the following is based on the *continuous version* of the Knapsack problem, it still is very useful and is solved independently of NEOS. Bobby Jaycox of the class of 2017 designed this App.

3.7.1 Introduction

The Knapsack problem (also referred to as the rucksack problem) is a classic CO problem. Simply put, there is a knapsack that can hold a maximum amount of weight. With this "knapsack" there are items that need to be put inside it. Given a set of items, each assigned a weight and value, the subject must determine the amount of each item that should be included in the collection. The results should provide a solution that makes the total weight inside the knapsack less than or equal to the allotted weight the knapsack can hold and makes the total value as large as possible.

3.7.2 Problem

The decision variables x_i of this linear program are the amount of an item the user will be allotting in the knapsack. Value variables a_i are given to items to measure how important they are while weight variables b_i are designated to each item to measure the total weight W that cannot be exceeded. The decision variables x_i have upper and lower bounds and are used to further define restrictions to each decision variable. This is a classical integer programming problem.

3.7.3 Mathematical Model

The decision variables of this linear program are the amount of the item the user will be allotting in the knapsack (x_i). Value variables $a_{ij}x_i$ are given to items to measure how prominent they are while weight variables (b_i) are designated to each item to measure the cost they will have to total weight that cannot be exceeded. The *At Most* variables (d_i) and *At Least* variables (c_i) are used to define restrictions to each decision variable.

$$Maximize\ Z = \sum_i a_i x_i \tag{3.46}$$

$$s.t. \sum b_i x_i \leq W\ \forall i \tag{3.47}$$

$$c_i \leq d_i\ \forall i \tag{3.48}$$

3.7.4 Algorithm

The algorithm examines the different combinations of items that do not violate the weight limit and maximize the value function. The logic of the algorithm is captured in the text below.

> Each specific statement tests to see which ratio is the greatest in value. If that ratio equals r_i, a second condition tests to see if x_i has reached its maximum value d_i. If it has not, then the "If" statement is entered. .01 is added to the value every time the "If" statement is entered to update the variables to get the best answer. Also, the nested "If" statement updates the ratio of the item as well. If the item has reached its capacity, the r_i is set to zero and the "Max_Ratio" variable is recalculated in the "Max_Ratio" procedure. x_i is then set to d_i.

The blocks run the program through the smartphone rather than through the NEOS server which would be necessary for larger more complex problems.

3.7.5 Solution App

Figure 3.18 illustrates the input requirements and how these input data are processed by the App. Figure 3.19 illustrates the blocks programming.

3.7.6 Evaluation

While the App does work, the problem is limited to only five items and larger numbers require the App to take excessive time to run. Even though it is not necessarily fast, the App provides optimal solutions for each variable as well as the optimal solution for the objective function. It can be used for a number of tasks and is practical for any situation where one might need to determine amounts of items to have given a limited weight to fit the items.

Another integer knapsack problem is described in the beginning of Chapter 4 which compliments this App.

3.8 Enumeration of Basic Feasible Solutions in Linear Programming

One way to solve some small Linear Programs is to enumerate all the basic solutions, then choose the feasible ones with the best objective. This is illustrative of the process of solving Linear Programs but can be very inefficient. It does not require any artificial variables and

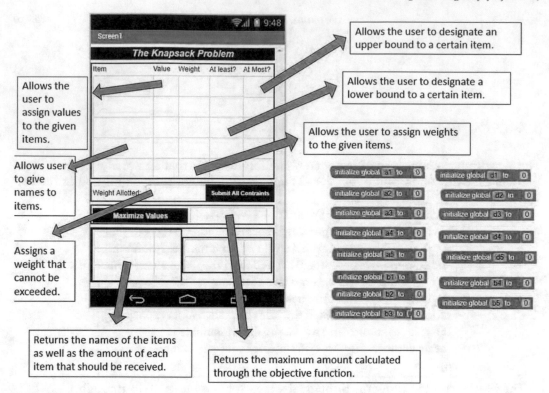

Fig. 3.18 Knapsack App

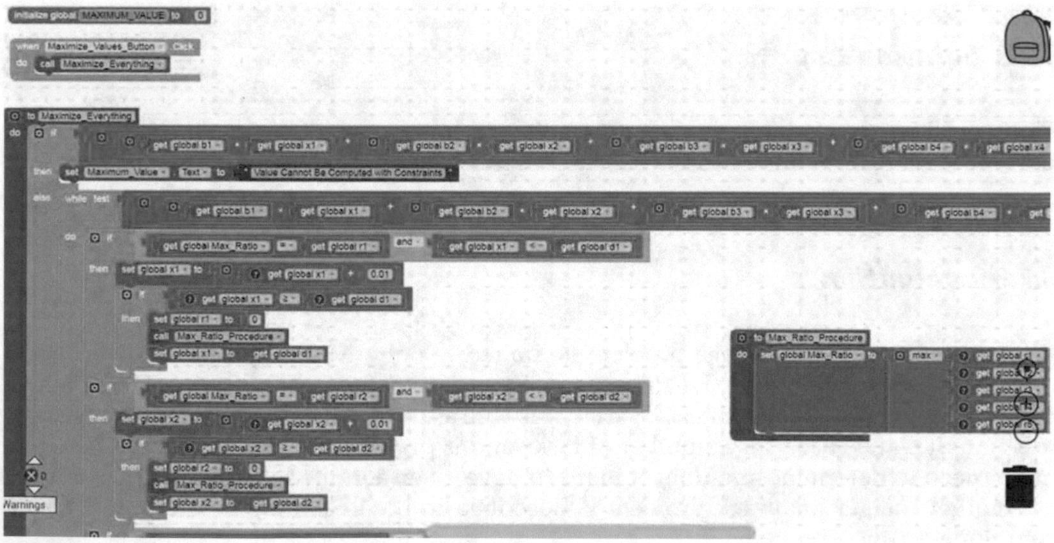

Fig. 3.19 Knapsack App

assumes that one is given a set of equations in a standard form where the number of rows is less than or equal to the number of columns. One may have to add slack or surplus variables in order to create the set of equations.

3.8.1 Introduction

Instead of a special algorithm for solving an LP, one can generate all the basic solutions, then choose the basic and feasible one which optimizes the objective function. This is of course very inefficient, but is a crude way of solving an LP without artificial variables or a complex pivoting process. For a phone App, it might be eminently suitable.

3.8.2 Problem

Let's posit some important definitions. A *feasible solution* is any nonnegative solution $x \geq 0$ such that the system of equations $Ax = b$ is satisfied. Without loss of generality, we can assume that the $m \times n$ matrix A has full row rank, so A has m linearly independent columns.

For some choice of m linearly independent columns of A called basic columns, a *basic solution* is any x such that $Ax = b$ and the $n - -m$ nonbasic variables are all zero. The basic variables then must be $x_B = B^1 b$ where the columns of B are the basic columns of A. Geometrically, a basic solution x_B defines an extreme point (a corner point) of the solution space $Ax = b$. Thus, searching all the corner points is a crude but potentially viable algorithm.

Thus, x_B is a basic solution if and only if B is nonsingular ($det B \neq 0$). If in addition, $x \geq 0$, then x is a *basic feasible solution (bfs)*.

Suppose we have the following system ($m < n$):

$$2x_1 + 3x_2 + x_3 = 5 \tag{3.49}$$
$$x_1 + 2x_2 + x_3 = 4 \tag{3.50}$$

$$\Rightarrow A = \begin{pmatrix} 2 & 3 & 1 \\ 1 & 2 & 1 \end{pmatrix}, b = \begin{pmatrix} 5 \\ 4 \end{pmatrix}$$

with $rank(A) = 2$, then

- **Solution: x** $= (-1, 2, 1)$
- **Feasible solution:** $(1/2, 1/2, 5/2)$
- **Basic solution:** $(-2, 3, 0)$
- **Basic feasible solution:** $(0, 1, 2)$

3.8.3 Mathematical Model

We are given a two-variable, two-row linear programming problem in standard equation form. We want to generate all the basic solutions to the system.

Since the number of rows is $m = 2$, one can use the following formulas for the determinant and basis inverse to construct the solution for the variables. Here are some useful formulas for computing the basic solution values. We need to compute the determinant of 2×2 matrices and then use it to compute the inverse and the solution for the two basic variables. Here are the formula:

$$A := \begin{pmatrix} a & b \\ c & d \end{pmatrix} \Rightarrow B := \begin{pmatrix} a & b \\ c & d \end{pmatrix}, \text{where } B^{-1} = \frac{1}{ad - bc} \begin{pmatrix} d & -b \\ -c & a \end{pmatrix}, \text{and} \Rightarrow det(D) := \frac{1}{ad - bc}$$

$$x = B^{-1}b \Rightarrow x_B = \begin{pmatrix} x_1 \\ x_2 \end{pmatrix} = \begin{pmatrix} dDb_1 - bDb_2 \\ -cDb_1 + aDb_2 \end{pmatrix}$$

3.8.4 Algorithm

A digression here, but it is important to point out that, a linear system in standard form in $n-$variables with rank m has a maximum of

$$\binom{n}{m}$$

basic solutions. In this case, the feasible ones can be found by setting $n-m$ of the variables at a time equal to zero, then solving, where possible for the remaining. This is generally an impractical approach, however, since

$$\binom{n}{m} = \frac{n!}{m!(n-m)!}$$

is usually a large number. Unfortunately this algorithm is *impractical, combinatorial, unbounded, and exponential* but it works pretty well in the smartphone environment. The App assumes that the problem is maximization.

A picture of the blocks used in the programming appears in Figure 3.20.

3.8.5 Demonstration

Let's determine the basic feasible solutions for the following LP.

$$\text{Max } z = 2x_1 - 4x_2 + 5x_3 - 6x_4 \tag{3.51}$$
$$\text{s.t.} \, x_1 + 4x_2 - 2x_3 + 8x_4 \leq 2 \tag{3.52}$$
$$-x_1 + 2x_2 + 3x_3 + 4x_4 \leq 1 \tag{3.53}$$
$$x_j \geq 0 \, \forall j \tag{3.54}$$

This is a nontrivial LP because it actually has six variables so we need a systematic procedure for its solution. Be sure to enter the slacks for the problem wherever appropriate since one is solving a set of equations. There are a total of fifteen basic solutions of which $x_1 = 8, x_3 = 3$ yields the optimal solution of $Z = 31$. The solution appears on the right screen in Figure 3.21.

3.8.6 Evaluation

The App works pretty well and it is especially useful to understand the notion of basic solutions and basic feasible solutions. Again, it is limited to $m = 2$ rows and a maximum of $n = 6$ variables. Any number from $n = \{2, 6\}$ decision variables can be used in the App. The LP is limited because the AI2 programming blocks were used to identify fifteen different basic feasible solutions. Of course the full App could be programmed to take advantage of the NEOS server to solve larger problems and that remains a good extension problem for future examination.

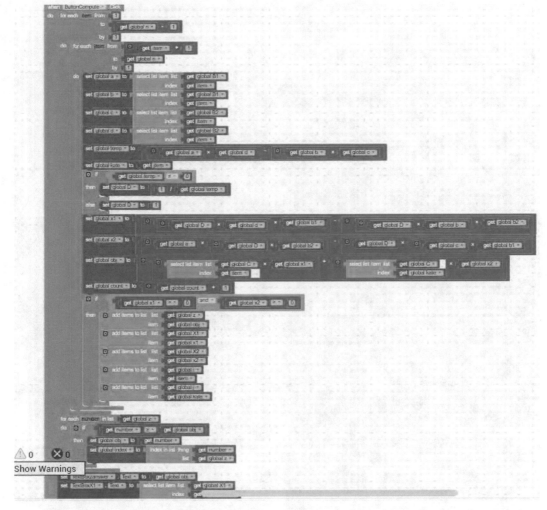

Fig. 3.20 Basic Feasible Solution Algorithm Blocks Programming

3.9 Simplex and Dual Simplex Algorithms App

3.9.1 Introduction

We use a special implementation of the AI2 programming language to demonstrate the simplex and dual simplex algorithms. We will only examine small problems with a few decision variables, but their implementation in AI2 is worthy of examination. The student who programmed AI2 is Allan Tang who was an undergraduate student in Computer Science at UMass in 2011, the first class in which I taught App Inventor. He programmed this initially in App Inventor Classic. It has been converted to AI2.

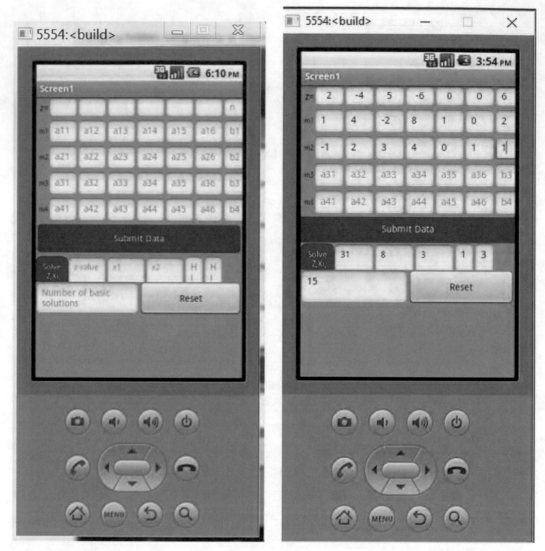

Fig. 3.21 Basic Solution App Programming and Sample BFS Solution

3.9.2 Problem

The problem for this App was to demonstrate the simplex algorithm and the dual simplex algorithm in an App setting. Allan Tang wanted to show that it was possible to solve small LPs with App Inventor and demonstrate the simplex tableaus which are challenging for students to understand.

3.9.3 Mathematical Model

One possible formulation of an LP is given below.

$$Maximize \ \ Z = \sum_j c_j x_j \tag{3.55}$$

$$subject \ to: \sum_j a_{ij} x_j \le b_i \ \forall i \tag{3.56}$$

$$x_j \ge 0 \ \forall j \tag{3.57}$$

This basic LP formulation of a maximization problem is what Allan utilized.

3.9.4 Algorithm

Figure 3.22 illustrates part of the blocks programming for the simplex App. It is a very efficient implementation.

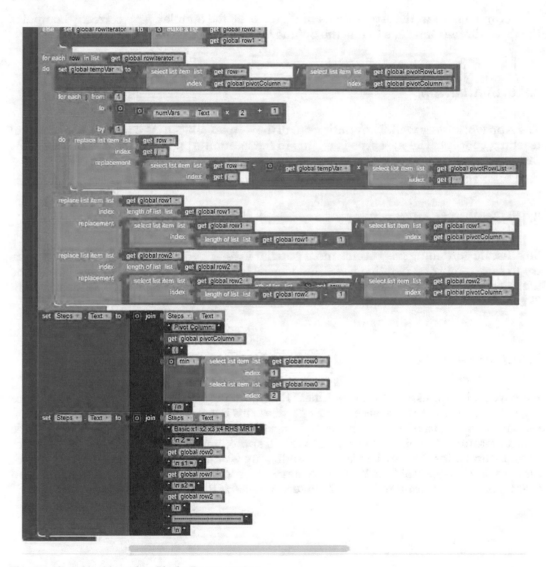

Fig. 3.22 Simplex Algorithm Blocks Programming

3.9.5 Demonstration

We will demonstrate a simple two-dimensional example. The App is capable of two- and three-dimensional examples with some nice interactive features to change the problems and parameters of the problems.

Let's say that we have the following two-dimensional LP.

$$\text{Maximize } Z = 350x_1 + 120x_2 \tag{3.58}$$

$$s.t. \quad 12x_1 + 5x_2 \leq 1850 \tag{3.59}$$

$$8x_1 + 2x_2 \leq 1200 \tag{3.60}$$

$$x_1, x_2 \geq 0 \tag{3.61}$$

The optimal solution is

$$x_1 = 143.75, x_2 = 25, Z = 5331. \tag{3.62}$$

As can be seen in the App screens of Figure 3.23, the Simplex App correctly computes the three simplex tableaus to find the optimal solution. This is quite a feat with AI2.

3.9.6 Evaluation

The App works very well. The printing out of the simplex tableau is an important and useful teaching concept. The next App is related to the location App just presented but concerns the detailed allocation of plants to a garden plot.

3.10 Garden Planting App

Small-scale gardening has become more popular in recent years as hobbyists and amateurs try to save money by planting herbs and vegetables in backyards and window boxes. The idea of the App is to assist gardeners in what to plant. Thomas Johnson a student in MIE 379 initially designed the App in 2014. Also, David Metz built upon Tom's App in 2017.

3.10.1 Introduction

Here we will demonstrate the use of Linear Programming (LP) to aid a gardener in determining what type of plants to place in their garden. This is a classic form of resource allocation planing with LP. In one sense this is an Integer Programming (IP) problem because of the integer nature of the planted items, but we will approximate this with continuous variables as an LP and allow for partial plants. Rounding up is also a viable approach for this type of resource allocation problem. We usually have a restriction on the total area A which we can plant along with a limiting budget B. Figure 3.24 starts the Garden App.

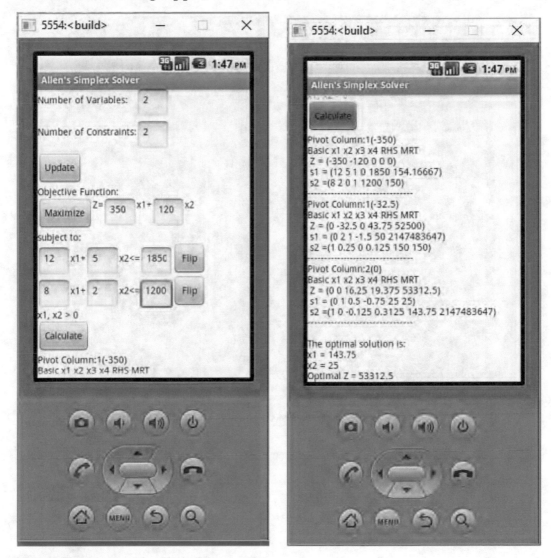

Fig. 3.23 Simplex Algorithm Blocks Programming

3.10.2 Problem

Choosing what to plant, then, is a nontrivial task. This App addresses that challenge using linear programming. It takes into account several pieces of information (it i.e. parameters) that have to be pre-specified:

(1) User's plant preferences, p_j
(2) Available space in the garden and the space that each plant requires $space_j, A$.
(3) Overall budget and the cost of each plant, c_j, B
(4) Area dedicated to each plant last year a_{j-1} is factored in order to balance the allocations a_j for the current year.

We assume that the gardener can either plant partial plants or start from seeds. Seeds are bought individually. We have information of the area requirements for each plant type. There are nine different decision variables or plant types:

Fig. 3.24 Garden Inputs/Outputs

- x_{BR} := Broccoli plants
- x_{LL} := Leaf Lettuce plants
- x_{Ce} := Cabbage plants
- x_{SC} := Swiss Chard plants
- x_{Pe} := Pepper plants

- x_{BB} := Bush Bean plants
- x_{Ct} := Carrot plants
- x_{Sp} := Spinach plants
- x_{Ra} := Radish plants

We could have more decision variables given the structure of the equations and constraints.

3.10.3 Mathematical Model

$$\text{Minimize } Z = \sum_{j=1}^{n} c_j x_j \tag{3.63}$$

$$s.t. \sum_{j=1} a_j x_j \leq A \qquad \text{area requirements} \tag{3.64}$$

$$\sum_{j=1} c_j x_j \leq B \qquad \text{budget constraint} \tag{3.65}$$

$$a_j x_j + a_{j-1} x_{j-1} \leq A \ (j = 1..n) \ \text{year} - \text{to} - \text{year limits} \tag{3.66}$$

$$x_j \geq 0 \ \forall j \ \text{plants} \tag{3.67}$$

where

- c_j:= Cost per item planted
- x_j:= Continuous variable for item planted
- a_j:= Area requirements of item to be planted
- A:= Area of the planting box;
- B:= The planting budget.

Figure 3.25 further illustrates graphically what the mathematical model is trying to accomplish.

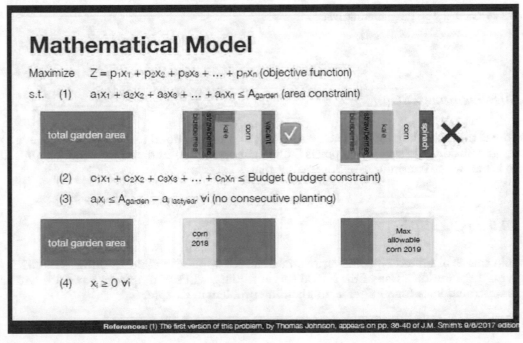

Fig. 3.25 Model Graphic

3.10.4 Algorithm

Figure 3.26 and 3.27 illustrate the programming blocks and its solution for the sample data. It works very well and is very fast. The bounds on the areas are critical in regulating the distribution of the plants. It is pretty straightforward as it relies on AMPL's mathematical programming language to encode the problem. Utilizing the mathematical programming language properties in AMPL can significantly simplify the problem formulation. While it is a bit abstract, once you learn the way to simplify the equations, it can make the blocks programming much more efficient.

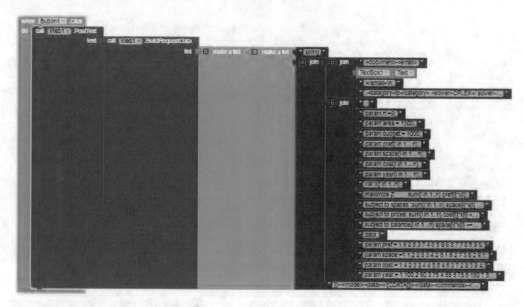

Fig. 3.26 Garden App Programming Blocks

3.10.5 Demonstration

Below is the continuous solution. One sees that we need to round up or down to achieve an integer solution. The second output is the integer optimal solution which is not obvious. It also is below the continuous optimum, but that is expected.

3.10.6 Evaluation

While this is a limited App, it demonstrates a nice application and usefulness of AI2 for solving LP solutions. Use of the NEOS server for larger LPs with integer variables will be demonstrated later for a larger farm planning problem in Chapter 5.

```
* * * * * * * * * * * * * * * * * * * * * * * * * * * * * * * * * * * * * * * * * * * * * * * * * * * *

        NEOS Server Version 6.0
        Job#      : 10027466
        Password  : NvauLPxk
        User      :
        Solver    : lp:bpmpd:AMPL
        Start     : 2021-01-10 15:45:37
        End       : 2021-01-10 15:45:43
        Host      : neos.la.asu.edu
* * * * * * * * * * * * * * * * * * * * * * * * * * * * * * * * * * * * * * * * * * * * * * * * * * * *
BPMPD 2.11: Optimal solution found, Z= 1428.364601
0 iterations, 0 corrections
x [*] :=
1  14.8515
2  28.8462
3  19.2308
4  26.3158
5  19.7368
6   9.86842
7  28.3019
8  16.3043
9  14.8515

BPMPD 2.11: Optimal integer solution found, objective Z= 1384
0 iterations, 0 corrections
x [*] :=
1  14
2  28
3  19
4  26
5  19
6   9
7  28
8  16
9  14
```

Fig. 3.27 Garden App Solution

3.11 Cardio Workout Coach

The following App presents a straightforward use of LP to select Cardio exercises. Erin O'Neil of the MIE 379 class programmed this App.

3.11.1 Introduction

Figure 3.28 illustrates the welcome screen and the first input screen for the preferences of the exercises.

People often struggle to correctly optimize a workout under time constraints. This App is designed to help people lose weight by burning a specific amount of calories to counteract the calories consumed. It is generally accepted that Cardio is the best way to burn calories, so the App is focused specifically on Cardio.

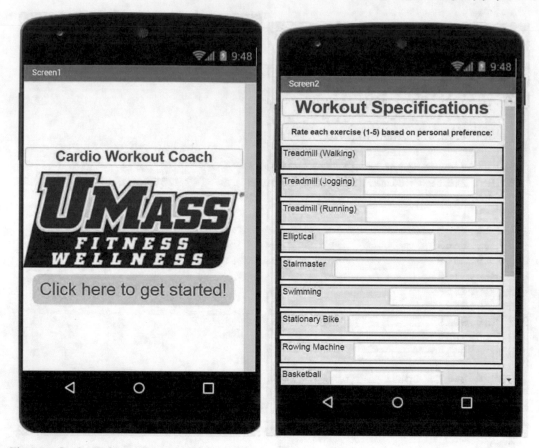

Fig. 3.28 Cardio Welcome Screen and Inputs

3.11.2 Problem

The App is modeled based on various exercises available at the UMass Recreation Center including walking, jogging, and running on the treadmill, using the elliptical, StairMaster, stationary bike and rowing machine, swimming, and playing basketball. For the purposes of this App, an "optimal workout" is defined as the workout which allows for maximum user satisfaction (based on input preferences) and meets the desired calorie goal.

3.11.3 Mathematical Model

Notation:

$$P_{1...9} := Workout\ Preferences\ entered\ by\ users.$$

$$x_{1...9} := Optimal\ time\ spent\ at\ each\ exercise.$$

$$x_{10...18} := Calories\ consumed\ at\ each\ exercise.$$

Assumptions:

- Calorie calculation equations from www.Livestrong.com
- Weight assumed to be weight of average American (CDC)

- Metabolic Equivalent Values (MET) information (National Cancer Institute & Institute of Lifestyle Medicine)

Calories per exercise

$$= \frac{Weight}{2.2046} * (MET\ Value) * \frac{Time\ at\ each\ exercise}{60} \tag{3.68}$$

$$x_m = 75.297 * (MET\ Value) * \frac{x_n}{60} \tag{3.69}$$

$$75.297 * METValue * \frac{x_n}{60} - x_m = 0 \tag{3.70}$$

Decision Variables:

$$x_{1\ldots9}\ \text{time spent at each exercise.}$$

Constraints:

$$6.275x_1 - x_{10} \leq 0 \tag{3.71}$$
$$10.043x_2 - x_{11} \leq 0 \tag{3.72}$$
$$14.425x_3 - x_{12} \leq 0 \tag{3.73}$$
$$11.288x_4 - x_{13} \leq 0 \tag{3.74}$$
$$10.043x_5 - x_{14} \leq 0 \tag{3.75}$$
$$12.550x_7 - x_{16} \leq 0 \tag{3.76}$$
$$5.827x_8 - x_{17} \leq 0 \tag{3.77}$$
$$10.043x_9 - x_{18} \leq 0 \tag{3.78}$$
$$x_1 + x_2 + x_3 + x_4 + x_5 + x_6 x_7 + x_8 + x_9 \leq Time\ Entered \tag{3.79}$$
$$x_{10} + x_{11} + x_{12} + x_{13} + x_{14} + x_{15} + x_{16} + x_{17} + x_{18} \geq Calories\ Entered \tag{3.80}$$
$$x_{1\ldots18} \geq 0 \tag{3.81}$$

Objective Function:

$$Maximize\ Z = P_1 x_1 + P_2 x_2 + P_3 x_3 + P_4 x_4 + P_5 x_5 + P_6 x_6 + P_7 x_7 + P_8 x_8 + P_9 x_9 \tag{3.82}$$

3.11.4 Algorithm

The NEOS server was used to solve the LP problem because of the large number of constraints. Figure 3.29 illustrates the Cardio blocks.

3.11.5 Solution App

Experimental Example Results
Objection function with hypothetical preferences:

$$Maximize\ Z = 3x_1 + 5x_2 + 3x_3 + 6x_4 + 2x_5 + 7x_6 + 8x_7 + 8x_8 + 4x_9 \tag{3.83}$$

- Time Available: 60 minutes
- Calorie Goal: 1000 calories

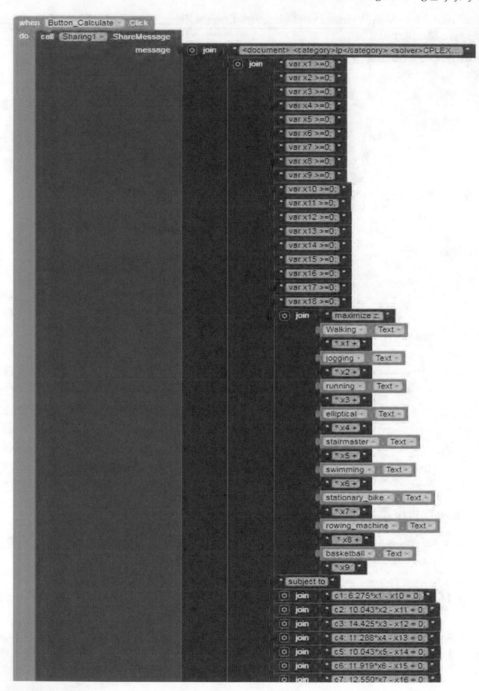

Fig. 3.29 Cardio Blocks

- Optimal Workout: $Z = 480, x_3 = 60$ *minutes* *Note x_3 has both high user preference and high MET value

  Figure 3.30 illustrates the NEOS solution for the Cardio App.

You are using the solver cplexamp.
Checking ampl.mod for cplex_options...
Checking ampl.com for cplex_options...
Executing AMPL.
processing data.
processing commands.
Executing on prod-exec-4.neos-server.org

18 variables, all linear
11 constraints, all linear; 36 nonzeros
        11 inequality constraints
1 linear objective; 9 nonzeros.

CPLEX 12.10.0.0: threads=4
CPLEX 12.10.0.0: optimal solution; objective 480
0 dual simplex iterations (0 in phase I)
x1 = 0
x2 = 0
x3 = 0
x4 = 0
x5 = 0
x6 = 0
x7 = 60
x8 = 0
x9 = 0
z = 480

Fig. 3.30 Cardio Solution from NEOS

3.11.6 Evaluation

The App is fairly straightforward and nicely utilizes the preference information to generate the objective function of the problem. In the next App, we show how LP can be used to solve a knapsack type of problem of special interest for outdoor enthusiasts.

3.12 Hiking App

Bekah Perlin programmed this App which uses the NEOS server to find the items needed on a hiking trip. It is similar to another type of Knapsack problem to be discussed in the beginning of Chapter 4. Figure 3.31 illustrates the Input screen.

3.12.1 Introduction

Hiking is one of the most popular outdoor activities, and for a good reason! As popular as it is, many people do not know how to appropriately pack food and water, sometimes packing more than needed (weighing themselves down), or not packing enough (returning hungry and dehydrated). This App will help beginners determine how much of each item they should bring. The hiker inputs their weight, distance of the hike, change in elevation, and six popular energy packed foods which are then rated (one or two depending on personal preference).

The output tells the user how many grams of each food they should bring based upon the fat, protein, and carbohydrate levels in each, as well as their preferences ratings of each. The user is also informed how many liters of water to bring, how much all of this should weigh, and how much space it will take in their pack. The user can take this information, and proceed on their adventure, stress free!

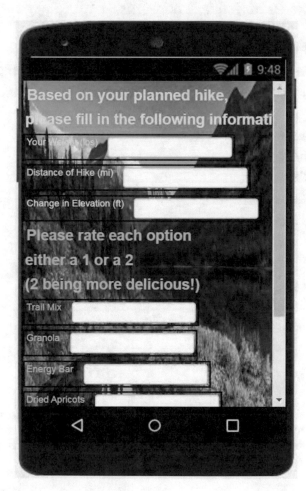

Fig. 3.31 Hiking Input

3.12.2 Problem

We need some backup data to help program the App. Table 3.1 illustrates the type of data collected for the Hiking App.

Data Collected

	Granola	Trail Mix	Energy Bar	Dried Banana	Dried Cranberry	Dried a Apricots
User Rating	R_G	R_T	R_E	R_B	R_C	R_A
Fat (%)	0.1	0.3	0.074	0.02	0.025	0.008
Protein (%)	0.1	0.133	0.147	0.04	0	0.031
Carb (%)	0.633	0.467	0.632	0.88	0.825	0.623
Weight per Volume (g/cup)	132	120	204	100	121	130
Volume per Weight (liter/g)	0.001792	0.001972	0.00116	0.002366	0.001955	0.018199

Table 3.1 Hiking Data

3.12.3 Mathematical Model

The model is basically a multi-item knapsack problem. Since we do not require integer variables, it can be solved with Linear Programming. We shall treat integer programming problems in the next chapter.

Relevant Information (assumptions):

- Calories per gram of each:
- Fat = 9 calories/g,
- Protein = 4 calories/g,
- Carbohydrate = 4
- Calories/gram, some of each food is required

Mathematical Model: Decision variables: $x_G, x_T, x_E, x_B, x_C,$ and x_A are the number of grams recommended of granola, trail mix, energy bar, dried banana, cranberry, and apricots, respectively.

Objective Function:

$$\text{Maximize } z : ((RG * x_G) + (RT * x_T) + (RE * x_E) + (RB * x_B) + (RC * x_C) + (RA * x_A)) \tag{3.84}$$

$$s.t. : (x_G * 0.1 * 9) + (x_T * 0.3 * 9) + (x_E * 0.07 * 9) + (x_B * 0.02 * 9) + (x_C * 0.03 * 9) +$$

$$(x_A * 0.008 * 9) \leq (\tfrac{1}{4}) * Chike \tag{3.85}$$

$$(x_G * 0.1 * 4) + (x_T * 0.13 * 4) + (x_E * 0.15 * 4) + (x_B * 0.04 * 4) + (x_C * .08 * 4) +$$

$$(x_A * 0.03 * 4) \leq (\tfrac{1}{4}) * Chike \tag{3.86}$$

$$(x_G * 0.63 * 4) + (x_T * 0.47 * 4) + (x_E * 0.63 * 4) + (x_B * 0.88 * 4) + (x_C * 0.83 * 4) +$$

$$(x_A * 0.62 * 4) \leq (\tfrac{1}{2}) * Chike \tag{3.87}$$

$$(x_G * 132) + (x_T * 120) + (x_E * 204) + (x_B * 100) + (x_C * 121) + (x_A * 130) \leq 23247 * d; \tag{3.88}$$

$$(x_G * CG * 0.001792) + (x_T * CT * 0.001972) + (x_E * CE * 0.00116) +$$

$$(x_B * CB * 0.002366) + (x_C * CC * 0.001955) + (x_A * CA * 0.02) <= 50 * Chike \tag{3.89}$$

3.12.4 Algorithm

Figure 3.32 illustrates the programming blocks. It employs the LP CPLEX algorithm on NEOS. The constraints on this problem are quite complex with all the nutrition requirements.

3.12.5 Solution App

Although the solution does not appear in the App, from the server email, we find the following. **Experimental Results: Inputs:**

- *Weight=165 lbs, Distance of the Hike=3 mi, Change in Elevation=500 ft.*
- *Ratings: Trail Mix =2; Granola =1; Energy Bar=2; Apricots =2; Bananas=2; Cranberries=2.*
- *Chike = 7171 calories.*

Outputs:

- $x_T = 177.592, x_G = 100, x_E = 68, x_A = 125, x_B = 150,$ and $x_C = 302.495,$
- *Water=4.916 Liters; Total Weight=3.92 lbs;* **Z=1596.18**

If we arrange for the variables to be integer, then

- $x_T = 177, x_G = 100, x_E = 68, x_A = 125, x_B = 150,$ and $x_C = 303,$
- *Water=4.68 Liters; Total Weight=3.9206 lbs;* **Z=1596.0**

This is a very similar solution to the continuous problem formulation.

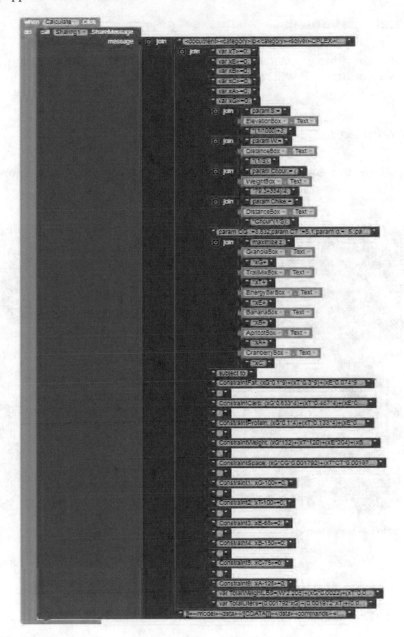

Fig. 3.32 Hiking Blocks

3.12.6 Evaluation

The App is nicely designed and quite complete in its input data requirements. In the next section of the book, the App is designed to utilize the NEOS server to solve general LP problems. Once the student understands the structure of the App, it can be expanded in the number of decision variables and constraints.

3.13 General LP Problem

Finally, we present a more general version of the LP which can be tailored to whatever application is appropriate. It will require the NEOS server, but it is generally foolproof.

3.13.1 Introduction

In order to solve some general LPs, we have the following input screen in Figure 3.33.

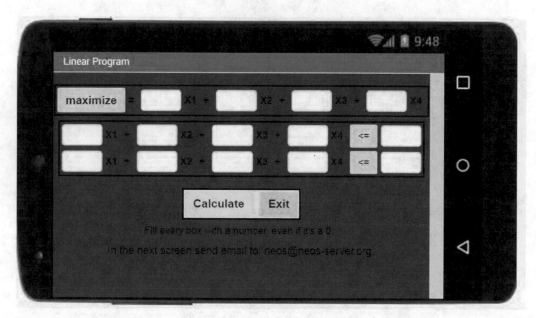

Fig. 3.33 Illustrate the Input Screen

As you can see. the App is set up for a maximization problem subject to two inequalities.

3.13.2 Problem

The App is flexible enough so that you can change the objective function from Maximum to Minimum or change the the inequalities from $\leq$ to $\geq$ just by pushing the inequality sign on the App. Make sure that you put data in all the cells of the model even zeroes. For instance, these blocks shown in the Figure show that we can treat the objective function and constraints and alter them to fit our model; please see Figure 3.34 and 3.35.

3.13.3 Mathematical Model

Our problem is as follows:

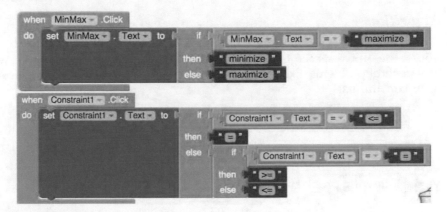

Fig. 3.34 Inequality Blocks

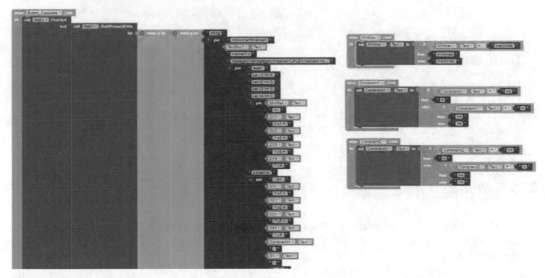

Fig. 3.35 General Block Structure

$$Maximize \ Z = \sum_j c_j x_j \tag{3.90}$$

$$subject \ to : \sum_j a_{ij} x_j \le b_i \ \forall i \tag{3.91}$$

$$x_j \ge 0 \ \forall j \tag{3.92}$$

If you examine the block structure and commands, it will solve the problem with CPLEX on the NEOS server. More information on the encoding in the blocks of the model of the problem is in the Appendix.

3.13.4 Algorithm

It does require the NEOS server for a solution delivered through an email, but it will also handle very general problems. Extensions could be generated for additional decision variables and/or constraints.

3.13.5 Solution App

Let's solve the following problem:

$$Maximize\ Z = 9x_1 + 12x_2 + 6x_3 + 4x_4 \tag{3.93}$$
$$subject\ to: 5x_1 + 8x_2 + 6x_3 + 3x_4 \leq 68 \tag{3.94}$$
$$x_1 - 2x_2 + 3x_3 - 4x_4 \geq 26 \tag{3.95}$$
$$x_j \geq 0 \, \forall j \tag{3.96}$$

The optimal solution returned by the server is

$$Z = 89.33: x_1 = 5.33, x_2 = 0, x_3 = 6.89, x_4 = 0 \tag{3.97}$$

We will follow up on this example in the next chapter when the variables should be integer/binary.

3.13.6 Evaluation

This shows that App Inventor together with the NEOS server works very well and the limits of the App Inventor program can be extended with the NEOS algorithms. This is quite promising.

Now we transition to the next chapter on Integer Programming (IP). Integer programming is an important and challenging research and teaching area. It is described in Chapter 9 of Taha's textbook, [14]. Taha does an excellent job of illustrating how integer programs are formulated and solved. Integer Programming methods are quite useful although difficult to solve in general since they fall into the realm of $\mathcal{NP}$-complete and $\mathcal{NP}$-Hard problems. Thus, it will be very difficult to verify optimality for large problem instances.

3.14 Exercises

1. Solve the following two problems with App Inventor:

   - i)

$$Minimize f(\bar{x}) = 3x_1 + 2x_2 + 7x_3 \tag{3.98}$$
$$subject\ to : 1x_1 + x_2 + x_3 \geq 6 \tag{3.99}$$
$$3x_1 - x_2 - x_3 \geq 1 \tag{3.100}$$
$$x_1, x_2, x_3 \geq 0 \tag{3.101}$$

   - ii)

$$Minimize f(\bar{x}) = 2x_1 - 2x_2 + 4x_3 \tag{3.102}$$
$$subject\ to : x_1 - x_2 + x_3 \leq 1 \tag{3.103}$$
$$x_1 + 2x_2 - x_3 \geq 6 \tag{3.104}$$
$$x_1, x_2, x_3 \geq 0 \tag{3.105}$$

2. **Diet LP Problems**

   Below is a sample of food items from the USDA[4] with nutrition information. Please develop a diet for these foods or others from the USDA website. Please add the *cost data* for the foods you selected.

Name	Calories	Cholesterol	Fat	Sodium	Carbo.	Fiber	Protein	Vit A	Vit C	Calcium	Iron
Frozen Broccoli	73.8	0.0	0.8	68.2	13.6	8.5	8.0	5867.4	160.2	159.0	2.3
Carrots, Raw	23.7	0.0	0.1	19.2	5.6	1.6	0.6	15471.0	5.1	14.9	0.3
Celery, Raw	6.4	0.0	0.1	34.8	1.5	0.7	0.3	53.6	2.8	16.0	0.2
Frozen Corn	72.2	0.0	0.6	2.5	17.1	2.0	2.5	106.6	5.2	3.3	0.3
Butter	35.8	10.9	4.1	41.3	0.0	0.0	0.0	152.9	0.0	1.2	0.0
Cheddar Cheese	112.7	29.4	9.3	173.7	0.4	0.0	7.0	296.5	0.0	202.0	0.2
Potatoes, Baked	171.5	0.0	0.2	15.2	39.9	3.2	3.7	0.0	15.6	22.7	4.3
Tofu	88.2	0.0	5.5	8.1	2.2	1.4	9.4	98.6	0.1	121.8	6.2
Roasted Chicken	277.4	129.9	10.8	125.6	0.0	0.0	42.2	77.4	0.0	21.9	1.8
Spaghetti W/Sauce	358.2	0.0	12.3	1237.1	58.3	11.6	8.2	3055.2	27.9	80.2	2.3
Scrambled Eggs	99.6	211.2	7.3	168.0	1.3	0.0	6.7	409.2	0.1	42.6	0.7
Bologna, Turkey	56.4	28.1	4.3	248.9	0.3	0.0	3.9	0.0	0.0	23.8	0.4
Banana	104.9	0.0	0.5	1.1	26.7	2.7	1.2	92.3	10.4	6.8	0.4
Grapes	15.1	0.0	0.1	0.5	4.1	0.2	0.2	24.0	1.0	3.4	0.1
Kiwifruit,	46.4	0.0	0.3	3.8	11.3	2.6	0.8	133.0	74.5	19.8	0.3

---

[4] US Department of Agriculture National Nutrient Database for Standard Reference.

3. **Investment Problem:**
   With the following definitions, please generate a Linear Program and find its solution. Graphically solve the problem then using the Simplex App

   $x_1 :=$ amount allocated to type X

   $x_2 :=$ amount allocated to type Y.

   ▸ The units will be in \$1000 chunks. For instance $x_1 = 50$ represents \$50,000 in type X. Since x_1 is an unknown decision variable, it must be established.

   ▸ Every \$100 invested in X returns annually \$5 translated into the units of the decision variable. This means that every unit of investment in type X of x_1 unit yields $\$50 \times x_1$ or $\$50 x_1$ dollars. If say $x_1 = 40$, this amounts to

   $$\$50 \times 40 = \$2000 \; x_1 \text{ total dollars}$$

   ▸ For every x_2 unit investment in Y of x_2 units, it return $\$80 \times x_2$ of $80 x_2$ dollars. Since we assume there is no interaction between the investments, the overall profit is

   $$50 x_1 + 80 x_2 = Z \text{ total dollars}$$

   of this chapter, solve the problem for the optimal investment.

4. **Knapsack Problem:**

   a) Solve the following (knapsack) problem:

   $$\text{Maximize } 2x_1 + 3x_2 + 8x_3 + x_4 + 5x_5 \qquad (3.106)$$
   $$\text{s.t.} : \; 3x_1 + 7x_2 + 12x_3 + 2x_4 + 7x_5 \leq 10 \qquad (3.107)$$
   $$x_1, x_2, x_3, x_4, x_5 \geq 0 \qquad (3.108)$$

5. **Knapsack Problem:**
   Based upon the Knapsack problem, develop a multi-constrained App, say with two separate resources, for example (*e.g.* Budget, Time, and Square Area Footage), for solving a knapsack problem on the NEOS server.

6. **Production Planning**

   A manufacturing firm would like to plan its production/inventory policy for the months of August, September, October, and November. The product under consideration is seasonal, and its demand over the particular months is estimated to be 500, 600, 800, *and* 1200 units, respectively. Presently the monthly production capacity is 600 units with a unit cost of $25. Management has decided to install a new production system with a monthly capacity of 1100 units at a unit cost of $30. However, the new system cannot be installed until the middle of November. Assume that the starting inventory is 250 units and that at most 400 units can be stored during any given month. Assume that the demand must be satisfied and that 100 units are required in inventory at the end of November.

   - **a)** *If the holding inventory cost per month per item is $3, find the production schedule that minimizes the total production and inventory cost using Linear Programming.*

7. **Trail Mix:**

   It is often hard to make a healthy trail mix. Most people tend to make trail mixes based on their favorite ingredients. This often leads to unhealthy mixes consisting mainly of sweets, such as chocolate and marshmallows instead of healthier ingredients such as nuts and raisins. Create a trail mix App that would blend the ingredients of different elements (*i.e.* nuts, oats, and granola) into a trail mix for hiking or other outdoor activities. Constraints would be for limiting the amount of sugar, fats, calories, cost, and so on as in Figure 3.36.

Name	Ingredient 1 Raisins	Ingredient 2 M&Ms	Ingredient 3 Peanuts	Ingredient 4 Cashews
Pref(1-10)	3	8	5	7
Calories	193	120	160	155
Sugar	11.7	15	1.3	2
Fat	9.3	5	14	12

Fig. 3.36 Trail Mix Example App

8. Garden Planting:

I have a small backyard garden that measures 10×20 feet. This coming spring I plan to plant three types of vegetable: tomatoes, green beans, and corn. The garden is organized in 10-foot rows. The corn and tomatoes are 2 feet wide and the beans are 3 feet wide. On a scale of $1 \rightarrow 10$, I would assign 10 to tomatoes, 7 to corn, and 3 to beans. My wife insists that I plant no more than two rows of tomatoes. How many rows of each vegetable should I plant?

▸ Formulate the garden planting problem as a Linear Program.

9. Blending Problem:

▸ Chandler Oil has 5000 barrels of Crude Oil 1 and 10,000 barrels of Crude Oil 2 available. Chandler sells gasoline and heating oil. These products are produced by blending together the two crude oils.

▸ Each barrel of Crude Oil 1 has an octane rating of 10 and each barrel of Crude Oil 2 has an octane rating of 5.

▸ Gasoline must have an *average* quality level of at least 8, while heating oil must have an *average* quality level of 6.

▸ Gas sells for $250/barrel and heating oil sells for $200/barrel. Advertising cost to sell one barrel of gas is $2.00 and the advertising cost of heating oil is $1.00. Let's assume that demand for heating oil and gas is unlimited so that all of Chandler's production can be sold.

▸ It is important to account for

  ▸ # of barrels of gas and heating oil produced.
  ▸ # of barrels of each crude oil used.
  ▸ quality level of inputs to make the outputs.
  ▸ total profits.

Integer Programming $\sum c_j \delta_j, \delta_j \in \{0,1\} \ \forall j$

Overview Integer Programming (IP) is one of the most useful optimization procedures because of its practical nature, however, it is also one of the most complex to solve because of the nonlinear, non-convex effects of the integer decision variables. Most of the IP problems for large scale instances are at least $\mathcal{NP}-$ Complete for the decision problem, and many are $\mathcal{NP}-$ Hard for the optimization part, so that getting optimal solutions is often out of the question.

For example: Maximize Z= $(550 * x_1 + 500 * x_2)$, subject to: $4 * x_1 + 5 * x_2 \leq 2000, 5 * x_1 + 4 * x_2 \leq 2200, 2.5 * x_1 + 7 * x_2 \leq 1750, x_1, x_2 \geq 0$ and integer. Optimal solution, $Z = 249,800, x_1 = 336, x_2 = 130$ (green dot in center). See Figure 4.1.

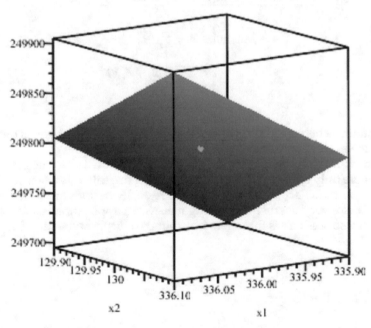

Fig. 4.1 Integer Programming Problem

Keywords: Integer Solutions, Unimodularity,

© Springer Nature Switzerland AG 2021
J. MacGregor Smith, *Combinatorial, Linear, Integer and Nonlinear Optimization Apps*,
Springer Optimization and Its Applications 175,
https://doi.org/10.1007/978-3-030-75801-1_4

God made the integers, man made the rest.

—Leopold Kronecker

The best thing about Boolean variables is even if you are wrong, you are off only by a bit

—Anonymous

If I wanted to solve an integer program, the first thing I would do is to solve the Linear Program (maximization) and "if the answer turned out to be $7\frac{1}{4}$ then I would at least know that the integer maximum could not be more than 7. No sooner had I made this obvious remark to myself that I felt a sudden tingling in my two left toes, and with great excitement realized that I had just done something different,"

—Ralph Gomory

The term 'totally unimodular' is due to Claude Berge, and far superior to our wishy washy phrase 'matrices with the unimodular property.' Claude had a flair with language.

—A. J. Hoffman

4.1 Introduction

As was discussed, Integer Programming (IP) problems have many potential applications since most real-world planning and design problems involve integer variables. It is a very rich area and very interesting and there is much research going on for the development of fast algorithms for large scale IPs.

- The general Linear Integer Programming (LIP) problem can be mathematically written as Find the integer variables $\mathbf{x} = \{x_1, x_2, \ldots x_n\}$ which

$$\text{Maximize } Z = \sum_j c_j x_j \tag{4.1}$$

$$s.t. \ : \sum_j a_{ij} x_j \leq b_i \ \forall i \tag{4.2}$$

$$x_j \geq 0 \ \forall j \tag{4.3}$$

- The only difference between this problem and the general LP problem is the requirement that x_j be integer, (*i.e.* the divisibility assumption no longer holds).
- The problems where all x_j are required to be integer are called *pure* integer, whereas those problems where only some of the x_j must be integer are called *mixed* integer problems.
- Problems where the x_j are binary variables (*zero-one*) variables are called *binary* integer programs. BIP are very useful in modeling real-world applications as we shall see.
- What we will also see here is that there is no simple simplex type algorithm for IPs (Fig. 4.1).

4.2 You Do Care Packaging

Based upon the Knapsack problems discussed in Chapter 3, we present an elaborate example to demonstrate what is possible with the App Inventor and NEOS.

4.2.1 Introduction

The basis of this project is a care package App that a person can use in order to optimally fill a flat shipping rate box from USPS with goodies. Originally this idea came from a common kindergarten activity, where kids make a greeting card to soldiers overseas thanking them. However, seeing that care packages are not limited to soldiers overseas, especially considering that parents often send care packages to their children in college, this is an App that anyone could use.

4.2.2 Problem

Objective: Fill a 11" × 8.5" × 5.5" flat rate box with as many objects as possible without going over the weight and size restrictions.

  Resources: For the purposes of this project, the resources were limited to 10 common objects that can be often found in care packages: Candy Bars, Five Gum, Altoid Mints, Cookie Packs, K-Cups, Single Serving Chips, Soda Cans, a Dollar Tea Can, a Beanie Baby, and a Greeting Card.

  Constraints: The main constraint in this linear program model is not exceeding the volume restrictions of the box. This is limited to the dimensions of the shipping box. There are also weight restrictions to the box as well, which is common in most shipping practices. A price section was also added to the care package, firstly to tally up the total price of the contents of the box, and secondly because price is a limiting factor in every realistic model. Finally what is inside the box is tailored to the preferences of the user. These choice are made by the user in the App.

  Other Assumptions: Other constraint assumptions were made in-order to simplify the model (Fig. 4.2).

- Out of the selected items, there should at least be one of each item.
- The maximum value the contents is thirty dollars.
- The dimensions of each item were assumed to be square in order to overestimate budget, giving leeway in volume consumption.
- Only one of each sentimental item is needed and required.
- Only one Beanie Baby and one Greeting Card can be included.
- If drinks are selected, the user can either send soda or a dollar tea, not both.

      1 Dollar Tea Can Limit
      2 Soda Can Limit
      5 K-Cups limit

The data used to estimate some of the parameters is shown in Figure 4.3.

4.2.3 Mathematical Model

The mathematical model is similar to that of the Knapsack Problem discussed in Chapter 3 except that we have integer variables (Fig. 4.4).

Fig. 4.2 Care Input Screen

4.2.4 Algorithm

The NEOS server was used because of the complex number of integer variables and constraints. A Branch and Bound algorithm was employed here.

Figure 4.5 illustrates the blocks programming used in the App.

Data

Items:	Height (in)	Width (in)	Length (in)	Weight (oz.)	Weight (lb.)	Volume (in^3)	Price/Unit ($)
Arizona/Peace Tea	7.5	3.8	3.2	16	1	91.2	$ 1.00
Soda Can	5.2	2.8	2.6	7.84	0.49	37.856	$ 1.50
Single Serving Chips	5	5	2	1	0.0625	50	$ 1.00
K-Cups	0.75	2	2	0.5	0.03125	3	$ 0.75
Oreo Snack Pack	5.5	3	2.5	1.74	0.10875	41.25	$ 1.50
Chips Ahoy! Snack Pack	5.5	3	2.5	1.74	0.10875	41.25	$ 1.50
Nutter Butter Snack Pack	5.5	3	2.5	1.74	0.10875	41.25	$ 1.50
Altoids	1	3.7	2.7	0.88	0.055	9.99	$ 1.50
Five Gum	0.6	3.3	3.3	1.76	0.11	6.534	$ 1.50
Hersey Candy Bar	0.4	6	2.5	1.55	0.096875	6	$ 1.50
Twix Candy Bar	0.6	4.5	1.9	1.79	0.111875	5.13	$ 1.50
Kit Kat Candy Bar	0.6	5.5	4	1.4	0.0875	13.2	$ 1.50
Reeces Candy Bar	0.75	6	4	1.5	0.09375	18	$ 1.50
Snickers Candy Bar	0.7	1.18	4	1.86	0.11625	3.304	$ 1.50
3 Musketeers Candy Bar	0.7	1.18	4.5	1.92	0.12	3.717	$ 1.50
Milky Way Candy Bar	0.7	1.18	4	1.82	0.11375	3.304	$ 1.50
Mounds Candy Bar	0.7	1.18	4	1.82	0.11375	3.304	$ 1.50
Almond Joy Candy Bar	0.7	1.18	4	1.82	0.11375	3.304	$ 1.50
Beanie Baby	5	2	9	4.5	0.28125	90	$ 7.50
Sun Glasses	6	2	1	1.176	0.0735	12	$ 1.30
Greeting Card	0 N/A	N/A		2	0.125	0	$ 3.00

USPS: Medium Flate Rate				Height (in)		11
				Width (in)		8.5
				Length (in)		5.5
				Volume (in^3)		514.25
				Weigth (lb)		70
				Price ($)	$	12.95

Fig. 4.3 Data

4.2.5 Solution App

A test was run using the App. The goodies that were put in the care package were candy bars, five gum, altoid mints, cookie packets, k-cups, single serving chips, a can of dollar tea, a beanie baby, and a greeting card.

The NEOS server responded with CPLEX 12.6.2.0: threads=4 CPLEX 12.6.2.0: optimal solution; objective 19.49101763 2 dual simplex iterations (0 in phase I)

$$x_1 = 3. x_2 = 1, x_3 = 1, x_4 = 1, x5 = 5, x6 = 4, x7 = 0, x8 = 1, x9 = 1, x_10 = 1, \text{Total Cost} = 41.2, z = 18 \quad (4.4)$$

This matches the initial example trial solution for the model above.

Ampl Code/LP Model

```
#--------Parameters--------    #----------Variables------------    #----------answer-------    #--------data------------
param candybar integer;        var x_1 integer >= candybar;       solve;                      data;
param candybarf integer;       var x_2 integer >= gum;            #---Nearest Integer ----     param candybar := 1;
param gum      integer;        var x_3 integer >= altoids;        let x_1 := floor(x_1);      param candybarf := 15;
param gumf     integer;        var x_4 integer >= cookies;        let x_3 :=floor(x_3);       param    gum := 1;
param altoids  integer;        var x_5 integer >= kcup;           let x_4 :=floor(x_4);       param    gumf := 15;
param altoidsf integer;        var x_6 integer >= chips;          let x_5 :=floor(x_5);       param altoids  := 0;
param cookies  integer;        var x_7 integer >= soda;           let x_6 :=floor(x_6);       param altoidsf := 0;
param cookiesf integer;        var x_8 integer >= tea;            let x_7 :=floor(x_7);       param cookies  := 1;
param kcup     integer;        var x_9 integer >= beanie;         let x_8 :=floor(x_8);       param cookiesf := 15;
param kcupf    integer;        var x_10 integer >= card;          let x_9 :=floor(x_9);       param  kcup  := 5;
param chips    integer;        var TotalCost integer >= 0;        let x_10:=floor(x_10);      param  kcupf := 5;
param chipsf   integer;                                                                      param  chips := 1;
param soda     integer;                                                                      param  chipsf := 15;
param sodaf    integer;        #------------Display Results---------------                    param   tea := 0;
param tea      integer;        display z;            display x_6;                            param   teaf :=0;
param teaf     integer;        display x_1;          display x_7;                            param  soda := 0;
param beanie   integer;        display x_2;          display x_8;                            param  sodaf := 0;
param beanief  integer;        display x_3;          display x_9;                            param beanie := 1;
param card     integer;        display x_4;          display x_10;                           param beanief := 1;
param cardf    integer;        display x_5;          display TotalCost;                      param  card := 1;
var TotalCost integer >= 0;                                                                   param  cardf := 1;
#--------------------Objective----------------------------
maximize z: x_1 + x_2 + x_3 + x_4 + x_5 + x_6 + x_7 + x_8 + x_9 + x_10;
#----------------------------Constraints----------------------------
subject to a: 3.304*x_1 + 6.534*x_2 + 9.99*x_3 + 41.25*x_4 + 3*x_5 + 50*x_6 + 37.856*x_7 + 91.2*x_8 +
90*x_9 + 0*x_10 <= 514.25;
subject to b: 0.11625*x_1 + 0.11*x_2 + 0.055*x_3 + 0.10875*x_4 + 0.03125*x_5 + 0.0625*x_6 + 0.49*x_7 +
x_8 + 0.28125*x_9 + 0.125*x_10 <= 70;
subject to c: 1.5*x_1 + 1.5*x_2 + 1.5*x_3 + 1.5*x_4 + 0.75*x_5 + x_6 + 1.5*x_7 + x_8 + 7.5*x_9 + 3*x_10 <= 30;
subject to d: 0*x_1 + 0*x_2 + 0*x_3 + 0*x_4 + 0*x_5 + 0*x_6 + x_7 + 0*x_8 + 0*x_9 + 0*x_10 <= sodaf;
subject to e: 0*x_1 + 0*x_2 + 0*x_3 + 0*x_4 + 0*x_5 + 0*x_6 + 0*x_7 + x_8 + 0*x_9 + 0*x_10 <= teaf;
subject to f: 0*x_1 + 0*x_2 + 0*x_3 + 0*x_4 + x_5 + 0*x_6 + 0*x_7 + 0*x_8 + 0*x_9 + 0*x_10 <= kcupf;
subject to g: 0*x_1 + 0*x_2 + 0*x_3 + 0*x_4 + 0*x_5 + x_6 + 0*x_7 + 0*x_8 + 0*x_9 + 0*x_10 <= chipsf;
subject to h: x_1 + 0*x_2 + 0*x_3 + 0*x_4 + 0*x_5 + 0*x_6 + 0*x_7 + 0*x_8 + 0*x_9 + 0*x_10 <= candybarf;
subject to i: 0*x_1 + x_2 + 0*x_3 + 0*x_4 + 0*x_5 + 0*x_6 + 0*x_7 + 0*x_8 + 0*x_9 + 0*x_10 <= gumf;
subject to j: 0*x_1 + 0*x_2 + x_3 + 0*x_4 + 0*x_5 + 0*x_6 + 0*x_7 + 0*x_8 + 0*x_9 + 0*x_10 <= altoidsf;
subject to k: 0*x_1 + 0*x_2 + 0*x_3 + x_4 + 0*x_5 + 0*x_6 + 0*x_7 + 0*x_8 + 0*x_9 + 0*x_10 <= cookiesf;
subject to l: 0*x_1 + 0*x_2 + 0*x_3 + 0*x_4 + 0*x_5 + 0*x_6 + 0*x_7 + 0*x_8 + x_9 + 0*x_10 <= beanief;
subject to m: 0*x_1 + 0*x_2 + 0*x_3 + 0*x_4 + 0*x_5 + 0*x_6 + 0*x_7 + 0*x_8 + 0*x_9 + x_10 <= cardf;
#--------------------------------TotalCost--------------------------------
let TotalCost :=1.5*x_1 + 1.5*x_2 + 1.5*x_3 + 1.5*x_4 + 0.75*x_5 + x_6 + 1.5*x_7 + x_8 + 7.5*x_9 + 3*x_10;
```

Fig. 4.4 Linear Integer Program

```
Presolve eliminates 11 constraints and 5 variables.
Adjusted problem:
5 variables, all linear
2 constraints, all linear; 10 nonzeros
        2 inequality constraints
1 linear objective; 5 nonzeros.

CPLEX 12.10.0.0: threads=4
CPLEX 12.10.0.0: optimal solution; objective 19.29641821
2 dual simplex iterations (0 in phase I)
x_1 = 1
x_2 = 1
```

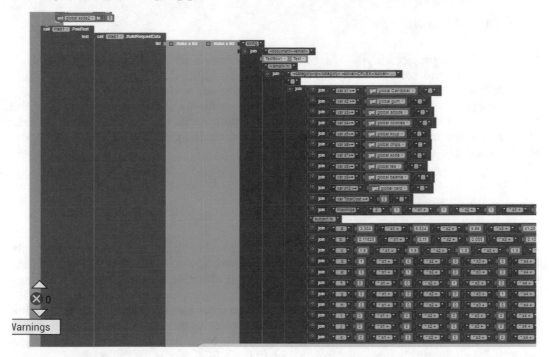

Fig. 4.5 Care Blocks Programming

```
x_3 = 1
x_4 = 1
x5 = 5
x6 = 5
x7 = 2
x8 = 0
x9 = 1
x_10 = 1
TotalCost = 41.2
z = 18
```

4.2.6 Evaluation (Benefits and Costs)

This is a valuable illustration of how AI2 can couple with the NEOS server to solve an interesting complex problem.

4.3 Project Scheduling App

This is a another classical example of an Mixed Integer Programming problem which is a classical Project Scheduling problem. The App was coded by Joe Woodman of the class of 2015.

4.3.1 Introduction

We have a list of project activities we need to schedule and we have a list of times for each project activity and certain due dates that need to be fulfilled. We are trying to schedule these activities for ourselves so there is only one processor or decision maker.

In most personal and professional settings, everyone needs to determine which activities to perform and how to schedule them over time. Being able to prioritize which projects should be completed and in what order can be most challenging when due dates, importance, and the time required to complete each task varies.

4.3.2 Problem

We would like to complete a set of activities over time where the due dates, project activity times, and preference importance for the activities are known. The App will create an estimated start date for each project activity. The overall objective is to minimize the lateness of the entire set of activities.

Figure 4.6 illustrates the process of inputting the data for the example demonstration where screen one on the left shows the start of the project and screen two on the right shows the data requirements for each project activity which must be input. Once all the project activity data are complete, one sends the App to the NEOS server for the solution.

4.3.3 Mathematical Model

The problem becomes a mixed-integer linear programming problem. The decision variables are

- $x_j :=$ the Start date for job j (measured from time zero)
-

$$y_{ij} = \begin{cases} 1 & \text{if } i \text{ precedes } j, \\ 0, & \text{if } j \text{ precedes } i. \end{cases} \tag{4.5}$$

The problem has two types of constraints: the noninterference constraints (guaranteeing that no two jobs are processed concurrently) and the due-date constraints. Two jobs i and j with processing time p_i and p_j will not be processed concurrently if (depending on whether which job is processed first)

$$x_i \geq x_j + p_j \text{ or } x_j \geq x_i + p_i \tag{4.6}$$

For M sufficiently large the "or" constraints are converted to "and" constraints by using

$$My_{ij} + (x_i - x_j) \geq p_j \text{ and } M(1 - y_{ij}) + (x_j - x_i) \geq p_i \tag{4.7}$$

The conversion guarantees that only one of the two constraints will be active at any one time.

Next, given that d_j is the due date for job j, the job is late if $x_j + p_j > d_j$. We use two nonnegative variables x_j^- and s_j^+ to determine the status of a completed job j with regard to its due date. The due date constraint is written as

$$x_j + p_j + s_j^- - s_j^+ = d_j \tag{4.8}$$

Job J is ahead of schedule if $s_j^- > 0$ and late if $s_j^+ > 0$. The mathematical model is then:

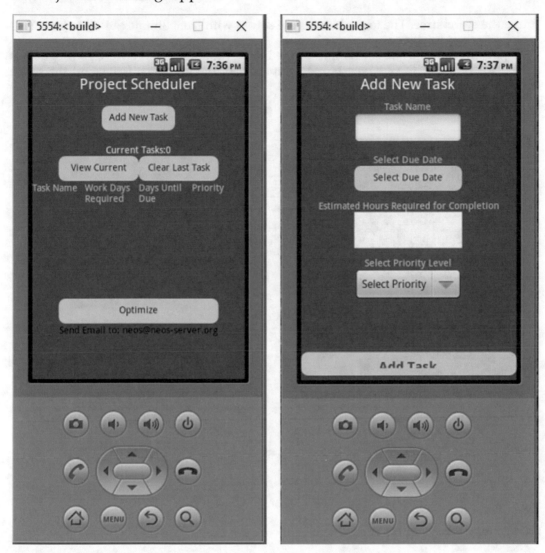

Fig. 4.6 Project Scheduling App Input Demo

$$\text{Minimize } Z = \sum w_j s_j^- \tag{4.9}$$

$$s.t. \; My_{ij} + (x_i - x_j) \geq p_j \tag{4.10}$$

$$M(1 - y_{ij}) + (x_j - x_i) \geq p_i \tag{4.11}$$

$$x_j + p_j + s_j^- - s_j^+ = d_j \tag{4.12}$$

$$x_j, s_j^-, s_j^+ \geq 0 \; \forall j \tag{4.13}$$

$$y_{ij} = (0,1) \forall (i,j) \tag{4.14}$$

Please see H. Taha's book [14], for example, in Chapter 9 for more details about this type of job scheduling model.

4.3.4 Algorithm

The underlying strategy for solving this problem is a *Branch & Bound* approach which is a classical method for solving IP problems. Linear Programming upper bounds are normally solved for at the beginning stages of the search process then lower bounds are estimated to prune the search tree. The lower bounds are solved with Dual Simplex LP iterations. This is also standard practice.

Figure 4.7 shows a partial sample of the blocks programming code. This program is solved for on the NEOS server as the complexity of the approach is beyond the scope of AI2.

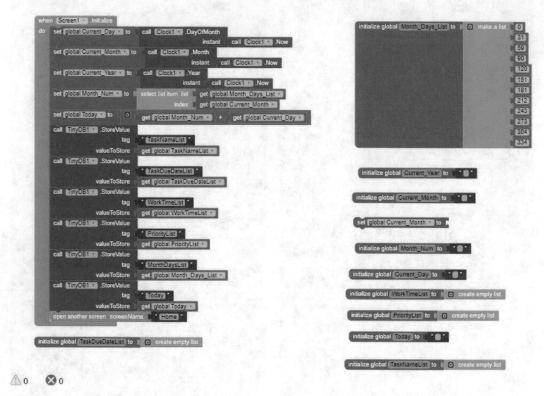

Fig. 4.7 Project Scheduling Blocks

4.3.5 Demonstration

The final solution from the NEOS server is for an example problem is

```
Presolve eliminates 0 constraints and 4 variables.
Adjusted problem:
24 variables:
        12 binary variables
        12 linear variables
28 constraints, all linear; 84 nonzeros
        4 equality constraints
        24 inequality constraints
1 linear objective; 4 nonzeros.
```

```
CPLEX 12.7.0.0: threads=4
CPLEX 12.7.0.0: optimal integer solution; objective 2.2312
42 MIP simplex iterations        0 branch-and-bound nodes
No basis.
penalty Z = 2.2312
x [*] := x_1=2.24   x_2= 0    x_3= 1.24    x_4= 0.62
```

4.3.6 Evaluation

The user interface is very good on this App. It is very easy to enter the project data.

4.4 Capital Budgeting

Capital Budgeting is a classical form of IP and has a great deal of historical experience. It involves the use of binary variables to help in selecting projects under usually stringent budgetary constraints. Tom Gutowski of the MIE 379 2017 class programmed the App.

4.4.1 Introduction

This App was developed to aid General Contractors in choosing between different projects. The revenue and annual cost of each project ultimately limits GCs from undertaking all possible projects. GC Project Selector allows a company to input the revenue, available funds and costs in order to choose the projects that will maximize profits while staying in budget over time.

4.4.2 Problem

This problem has similar properties to the multi-item knapsack problem. Figure 4.8 shows the input for the project selection process. The projects are examined over time for their profits subject to budgetary constraints over two years.

4.4.3 Mathematical Model

The form of the mathematical model is

$$\text{Maximize } Z = \sum_j c_j x_j \tag{4.15}$$

$$s.t. : \sum_j a_{ij} x_j \leq b_i \ \forall i \tag{4.16}$$

$$x_j (j \in (0,1)) \ \forall j \tag{4.17}$$

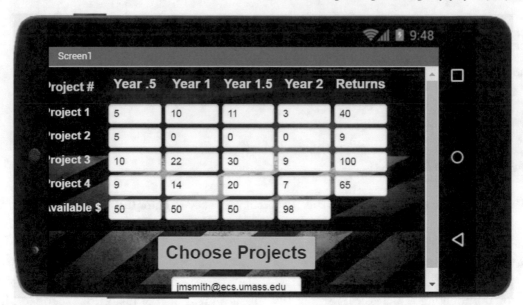

Fig. 4.8 GC Blocks Programming

4.4.4 Algorithm

Because of the binary variables, we will utilize the NEOS algorithm for linear-integer programming to solve the problem. Figure 4.9 illustrates the Blocks Programming.

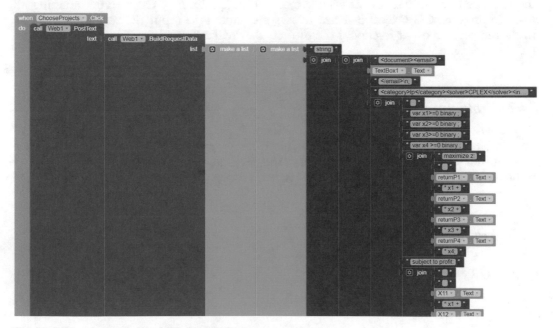

Fig. 4.9 Project Selection Blocks Programming

4.4.5 Solution App

Let's say we have the following 4-project problem:

$$Maximize\ Z : x_1 + x_2 + x_3 + x_4 \tag{4.18}$$
$$s.t. : x_1 + x_2 + x_3 + x_4 \leq x_{A1} \tag{4.19}$$
$$x_1 + x_2 + x_3 + x_4 \leq x_{A2} \tag{4.20}$$
$$x_1 + x_2 + x_3 + x_4 \leq x_{A3} \tag{4.21}$$
$$x_1 + x_2 + x_3 + x_4 \leq x_{A4} \tag{4.22}$$
$$x_j\ j \in (0,1)\ \forall j \tag{4.23}$$

Below is the solution of the problem.

```
You are using the solver cplexamp.
Checking ampl.mod for cplex_options...
Checking ampl.com for cplex_options...
Executing AMPL.
processing data.
processing commands.
Executing on prod-exec-4.neos-server.org

Presolve eliminates 3 constraints.
Adjusted problem:
4 variables, all binary
1 constraint, all linear; 3 nonzeros
        1 inequality constraint
1 linear objective; 4 nonzeros.

CPLEX 12.10.0.0: threads=4
CPLEX 12.10.0.0: optimal integer solution; objective 174
0 MIP simplex iterations
0 branch-and-bound nodes
x1 = 0
x2 = 1
x3 = 1
x4 = 1
z = 174
```

4.4.6 Evaluation (Benefits and Costs)

GC Project Selector:

- Maximizes profits
- Prevents exceeding the budget.

GC Project Selector ultimately allows GCs to meet client expectations by ensuring projects are completed on time and within the original job bid.

4.5 Machine Tool Selection

US Tsubaki, a roller chain manufacturing company located in Holyoke, MA must determine how many press punchers they should buy monthly for their press machines. There are five different types of punchers, each of these have different diameters and are used to punch holes through metal plates. This step counts as one of the first processes for the chain production process. Milagros Malo designed this App based upon demand, monthly budget, and price of each puncher.

4.5.1 Introduction

As these punchers punch through the metal plates, the metal to metal interaction (adhesive wear) increases, causing the metal to wear out and lose sharpness over time. With the multiple orders and demand of product to fulfill each month, Tsubaki needs to determine an accurate way to decide the number of each puncher size to have in stock at a given time. The App determines the output the optimal amount of punchers to purchase each month.

4.5.2 Problem

Historically, Tsubaki has placed monthly orders based on what a staff member can see available and unavailable in the puncher cabinet. This is not the best methodology because orders change monthly as the demand for certain products changes. Figure 4.10 illustrates the input data screens for the App.

Fig. 4.10 Tsubaki Input Screens

The App outputs the projected optimal number of each puncher type to purchase each month. It takes into consideration the price of each puncher, a monthly budget, product demand, and the utilization of each puncher.

4.5.3 Mathematical Model

$$Maximize\ Z = :22*x_1 + 20*x_2 + 20*x_3 + 18*x_4 + 14*x_5 \tag{4.24}$$
$$s.t.\ 22*x_1 + 20*x_2 + 20*x_3 + 18*x_4 + 14x_5; \le 75000 \tag{4.25}$$
$$x_1 + x_2 + x_3 + x_4 + x_5 \ge 50 \tag{4.26}$$
$$37000*x_1 + 47000*x_2 + 18400*x_3 + 25400*x_4 - 200*x_5 \ge 0 \tag{4.27}$$
$$700*x_1 + 700*x_2 - 300*x_3 - 300*x_4 - 300*x_5 \ge 0 \tag{4.28}$$

4.5.4 Algorithm

The Linear-integer programming algorithms of NEOS were used to solve the problem. Figure 4.11 illustrates the blocks used.

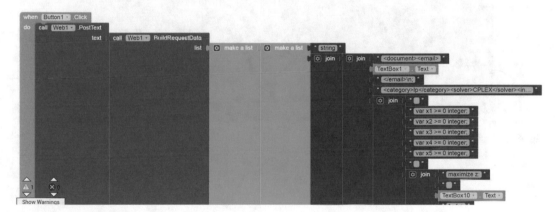

Fig. 4.11 Tsubaki Blocks

4.5.5 Solution App

For an example input of

- Puncher Type 1: 22
- Puncher Type II: 20
- Puncher Type III: 20
- Puncher Type IV: 18
- Puncher Type V: 14
- Number of Punchers: 50
- Budget $75000

Figure 4.12 illustrates the results from NEOS.

Fig. 4.12 Tsubaki Solution

4.5.6 Evaluation

For the example that Milagros used, the App gave an output of buying 15 punchers of Puncher 3 and 35 punchers of Puncher 5 for the coming month. Buying zero of other punchers is possible since some of the constraints are based on product demand and this can vary by month. It may also be possible that the other punchers were used less during this month and can still function for the next month. With the help of this App, one can project the minimum number of punchers to purchase each month.

As a future work on this App, one can try to continue to minimize the spending cost on these punchers by creating a system where an operator from the press room checks-out a puncher to be used and later checks it back into stock. The benefit to this idea is that it would lower the spending cost on punchers since many punchers go missing or are misplaced due to the lack of communication between shifts.

4.6 Restaurant Scheduling

Scheduling and sequencing problems are classic problems in Integer Programming. Andrew Giampa a member of the 2016 class worked part-time at a restaurant where scheduling the staff was critical for meeting the demand for customer dining. He created an App to help with staff scheduling during peak times at the restaurant.

4.6.1 Introduction

His restaurant model is based on dinner service hours between $5 and 10pm$; these operational hours are an average he found through research but also by his own experience in the restaurant business. The average amount of seats and tables for a restaurant in the U.S. is $120 - 140$ and $35 - 40$, respectively (quora.com). This was taken into account when researching for constraint statistics.

4.6.2 Problem

This essentially is an Integer Linear Program (ILP). The NEOS server must be utilized here to solve the problem.

4.6.3 Mathematical Model

The LP optimizes staffing according to the number of reservations scheduled on a night. The user simply inputs the amount of reservations per hour period, i.e. how many reservations between 6 and 7pm and so on. The idea behind this method comes from an average time of about an hour that customers will spend at the table. The *reservations* input could also be viewed as arrival rates per hour.

Two types of employees are scheduled for each job type. "All night" employees work the entire operational hours, 5-10pm, and "peak service" employees work from 6 to 9pm due to the major spike in reservations/arrival rates that tend to fall between these dinner hours.

4.6.4 Algorithm

The NEOS server LP with integer variables was used in the solution algorithm. *Objective function:* The goal of the objective function in this LP is to minimize labor costs, which are input by the user for each type of worker. Decision variables are $(x_n) n = 1 \ldots 4$ the amount of workers to schedule.

Constraints: Constraints for this LP correspond to demand rates for servers and cooks per hour during operational hours. Since there are 5 service hours, there are 10 constraints (5 for cooks and 5 for servers). The demand for each worker is calculated arithmetically in AI2 before the LP is solved.

Demand Calculations: A combination of research into restaurant statistics (quora.com) and personal knowledge of the business lead to worker constraint factors as follows:

Table-to-server ratio should be between 4 *and* 5. Three cooks can handle up to 20 reservations per hour, anything more than 20 requires an additional cook for up to 3 more reservations. This is to ensure an overload of reservations can still be handled within an hour.

4.6.5 Solution App

Figure 4.13 illustrates the blocks programming.

4.6.6 Evaluation

The restaurant staffing optimization App works, however many issues arose dealing with the emulator. It is ideal and easier to use an actual Android device in order to send the LP as an email. Also, this App only works for restaurants that only take reservations, and it does not account for walk-ins. However, if the amount of walk-ins to reservations is directly proportional, the theory behind the LP model will work.

4.7 Manufacturing Resource Constraint Problem

In a manufacturing process, there are four different tasks needed to be done. Each task costs different amount time and has different retail price. In order to maximize the total earning of the manufacturing, this App uses five inputs, unit prices of each task, and the total processing hours during a week. Johnny Zhu of the class of 2015 in MIE 379 programmed this App.

4.7.1 Introduction

This App will yield the result of the hours for each of the tasks every week with the maximum profitable earnings. The number of decision variables and constraints is not excessive. Since the results will be large numerical values, an Integer solution is sought.

Figure 4.14 illustrates the data inputs needed for running the App.

4.7.2 Mathematical Model

Target function:

$$Z = Input1 * X_1 + Input2 * X_2 + Input3 * X_3 + Input4 * x_4 \tag{4.29}$$

$$s.t.: 1.5 * x_1 + 1.1 * x_2 + 1.1 * x_3 + 0.55 * x_4 Input5 \tag{4.30}$$

$$x_1 \geq 3x_4 \tag{4.31}$$

$$x_2 \geq 1.5x_4 \tag{4.32}$$

$$x_3 \geq 1.5x_4 \tag{4.33}$$

$$12 \leq x_1 \leq 30, 12 \leq x_2 \leq 30, 20 \leq x_3 \leq 45, 5 \leq x_4 \leq 15 \tag{4.34}$$

- • Input1, Input2, Input3, Input4 are four unit prices
- • Input5 is the total processing hours

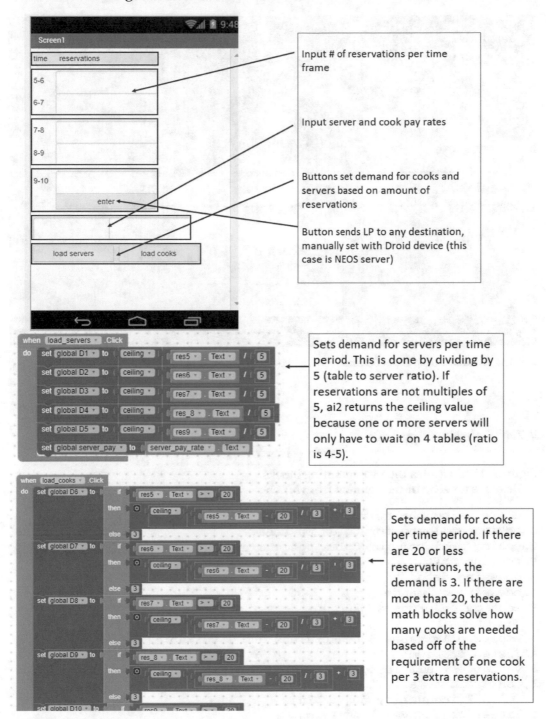

Fig. 4.13 Restaurant Scheduling App with Blocks Coding

4.7.3 Algorithm

We will use the Linear-Integer programming software from NEOS to solve this problem because of the integer variables.

Fig. 4.14 Manufacturing Input Screens

4.7.4 Solution App

Figure 4.15 illustrates the programming blocks.
    Below is an echo of the solution from NEOS.

```
You are using the solver Cplexamp.
Checking ampl.mod for Cplex_options...
Checking ampl.com for Cplex_options...
Executing AMPL.
processing data.
processing commands.
Executing on prod-exec-6.neos-server.org

Presolve eliminates 5 constraints and 1 variable.
Adjusted problem:
3 variables, all integer
3 constraints, all linear; 6 nonzeros
        3 inequality constraints
1 linear objective; 3 nonzeros.

CPLEX 12.10.0.0: threads=4
CPLEX 12.10.0.0: optimal integer solution; objective 2365
5 MIP simplex iterations
0 branch-and-bound nodes
x_1 = 20
x_2 = 34
x_3 = 40
```

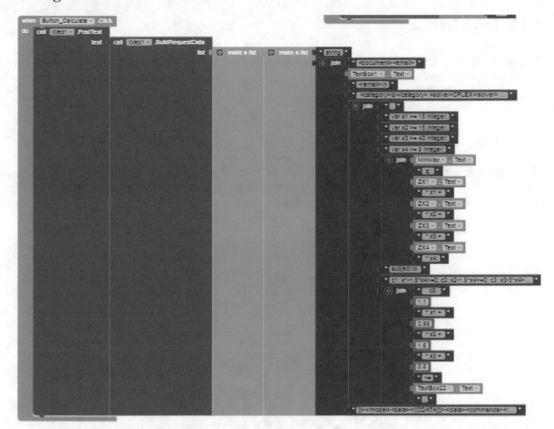

Fig. 4.15 Resource Planning Blocks Programming

```
x_4 = 13
z = 2365 <====================
```

While we did not run a series of different scenarios, different results would be calculated according to the different scenarios. Keeping the unit prices unchanged, different total working time leads to the different total profits. Further, the total profit increases linearly with the change of total working hours.

4.7.5 Evaluation (Benefits and Costs)

The App will successfully help factory foremen to design the basic process schedule depending on the four unit prices inputs and one total processes hours. Further upgrades would be focused on developing a user friendly interface and adding another four inputs for the unit time cost of the four processes.

4.8 Yoga Workout

This workout App was created by Rebecca Castonguay of the MIE 379 class of 2017. This App creates a yoga-based workout depending on the user's specific wants and needs. The App also shows the different types of exercises through a series of photos so that a new user can understand what the exercises names mean.

154 Integer Programming $\sum c_j \delta_j, \delta_j \in \{0,1\} \ \forall j$

4.8.1 Introduction

The user will input the duration and preference for each area of exercise and the App will give a written exercise plan for the user.

4.8.2 Problem

The App will set out to optimize the time given to create the best most rewarding workout possible. Figure 4.16 illustrates the first input screen for preferences. The use of the horizonal sliders is a very interesting way to input the preferences.

Fig. 4.16 Yoga Input Screen

4.8.3 Mathematical Model

The mathematical model of the problem appears in Figure 4.17.

Objective : $Maximize\ z = c_1x_1 + c_2x_2 + c_3x_3 + c_4x_4$

$z := user\ happiness$

$\Big\{$ $c_j := user\ input\ for\ desired\ workout\ preferences\ /\ j = 1,2,3,4$

$x_j := number\ of\ poses\ in\ each\ category\ /\ j = 1,2,3,4$

$\Big\{$ 1) *Flexibility* $:= x_1$

2) *Strength* $:= x_2$

3) *Balance* $:= x_3$

4) *Cardio* $:= x_4$

Notes : *Flexibility poses are 1 minute in length*

Strength poses are 45 seconds in length

Balance poses are 2 minutes in length

Cardio poses are 30 seconds in length

Constraints :

$(I)\ x_1, x_2, x_3 \geq 1$ → At least 1 of each pose is necessary

$(II)\ x_1 + .75x_2 + 2x_3 + .5x_4 \leq t$ → All activities must sum to equal the total workout length

$(III)\ .5x_4 \leq \frac{t}{6}$ → Total cardio cannot be more than ⅙ of total workout time

$(IV)\ x_3 \leq 4$ → No more than 4 balance poses allowed

$(V)\ x_1 \geq .75x_2$ → At least as much time must be spent on flexibility as time spent on strength

$(VI)\ x_4 \geq 0$ → Non-negativity constraint

Fig. 4.17 Mathematical Model

4.8.4 Algorithm

In this example, the user used the sliders in the App to input preferences of $1, 2, 5, 3$ for the *flexibility, strength, balance, and cardio* categories, respectively, and a total workout time of 30 minutes. This means that the user wanted a workout that emphasized balance and cardio and wanted minimal emphasis on flexibility. The App sent the data to the NEOS server, using AMPL as a template, and the NEOS server solved the linear program. There is no way for the server to connect back to Appinventor. As a result, the user will have to check their email for the optimal number of poses and then cross reference this number with the exercise category screen in the App.

With these user's preferences, the optimal workout would include 8 flexibility poses, 11 strength, 4 balance, and 10 cardio exercises. Using the exercise screens, the user can now choose which poses they would like to perform. As indicted on the screen, the user will perform each move for a certain time duration and then move immediately into their next pose of choice.

4.8.5 Solution App

Figure 4.18 illustrates the programming blocks while Figure 4.19 indicates the various exercises.

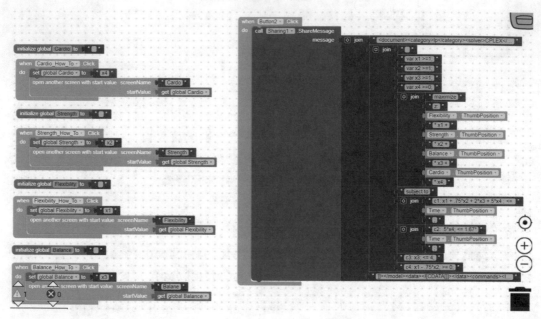

Fig. 4.18 Yoga App Blocks

Exercise Screens

Fig. 4.19 Yoga Exercise Screens

4.8.6 Evaluation

With people who have a desire to workout but do not know how to begin, the user is able to adjust their workout based on their preferences for that day. The constraints of the workout are based on the *Broga Yoga®method* of working out but they could be modified to fit any fitness format.

4.9 Newman Center Inventory Ordering

Sean Fitzgerald, class of 2016, worked at the Newman Center restaurant and wanted to develop an App for ordering items for the dining area which he supervised. This is an ambitious App given the number of variables and constraints employed but also reflective of a real problem.

4.9.1 Introduction

This is a classic inventory ordering problem. Vegetables and produce are critical components of the dining service and must be ordered weekly and there are many types of vegetables with varying demands.

4.9.2 Problem

There are many parameters and constraints in the problem so this is a good sized mixed MIP problem with integer and binary variables.

4.9.3 Mathematical Model

In the model, the objective function is to minimize the quantity ordered so as to keep spoilage to a minimum. There are a mix of continuous and integer variables so this is a mixed-integer programming problem. There are numerous constraints concerning the demand and the timing of deliveries during the ordering period. Figure 4.20 illustrates the mathematical model of the problem.

4.9.4 Algorithm

This is a large scale integer LP with many constraints, so it must be solved with the NEOS server.

4.9.5 Solution App

The overall number of variables and constraints of the model is quite impressive (Figs. 4.21, 4.22, and 4.23).

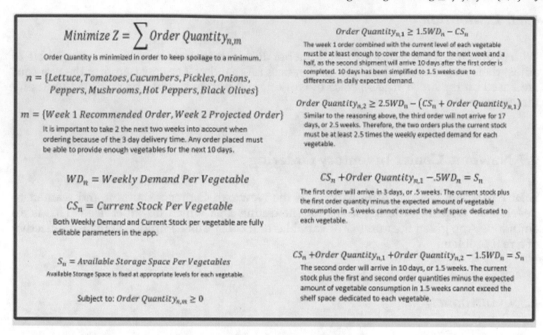

Fig. 4.20 Inventory Math Model

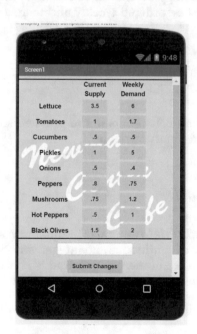

Fig. 4.21 Inventory App Input

4.9.6 Evaluation

Figure 4.24 illustrates the output from the NEOS server which is quite impressive.

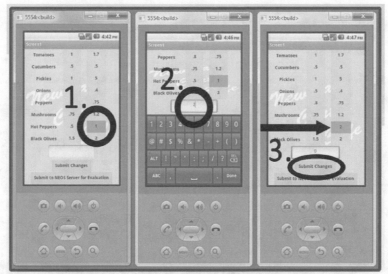

1. Select the stock or demand parameter you wish to change.
2. Input new value.
3. Submitted changes are recorded.

Fig. 4.22 Inventory App Input

4.10 Tournament Selection

This is a good example of scheduling for student activities. Rose Kelly, class of 2016, designed and programmed the App.

4.10.1 Introduction

This App is a complex one because of the number of decision variables. It also seeks to find an optimal integer solution.

This is an example of an integer program with binary decision variables which is going to be solved with the NEOS server. It represents a class of IP problem having to do with *Project Selection* which is a very valuable application in general.

4.10.2 Problem

A problem that is relevant to the college club sport community is determining in which tournaments to participate. Clubs, such as the Women's Ultimate Frisbee team, get invitations of upwards to twenty tournaments per semester, and realistically can only attend 4–6 of those. The objective of the App is to find the optimal combination of tournaments that maximizes team preference while abiding by the given constraints of budget and time duration. Each variable will be a binary $(0, 1)$ integer variable representing each individual tournament. The tournaments will have a preference on a scale from $1 \rightarrow 5$. Teammates will rate each

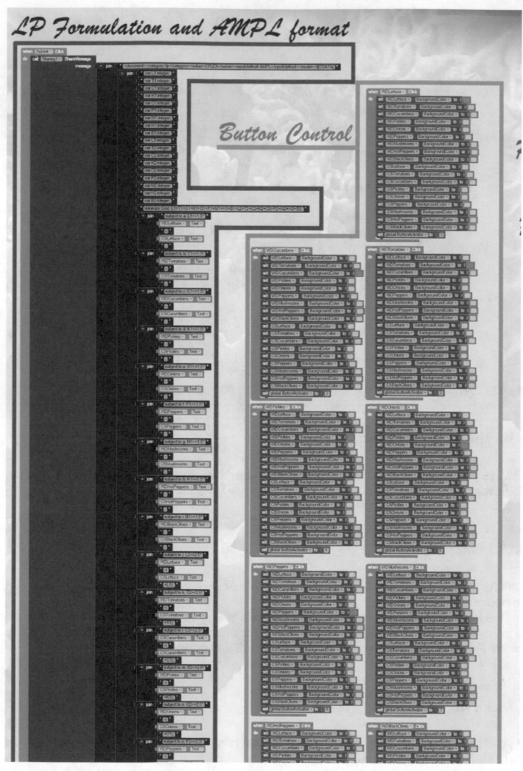

Fig. 4.23 Inventory App Blocks

tournament. Each tournament will have a cost and the constraints will be to remain under the selected total budget and time constraints. Also, because each tournament will be either a one-day or a two-day tournament, there will be restraints as far as how many two-day or

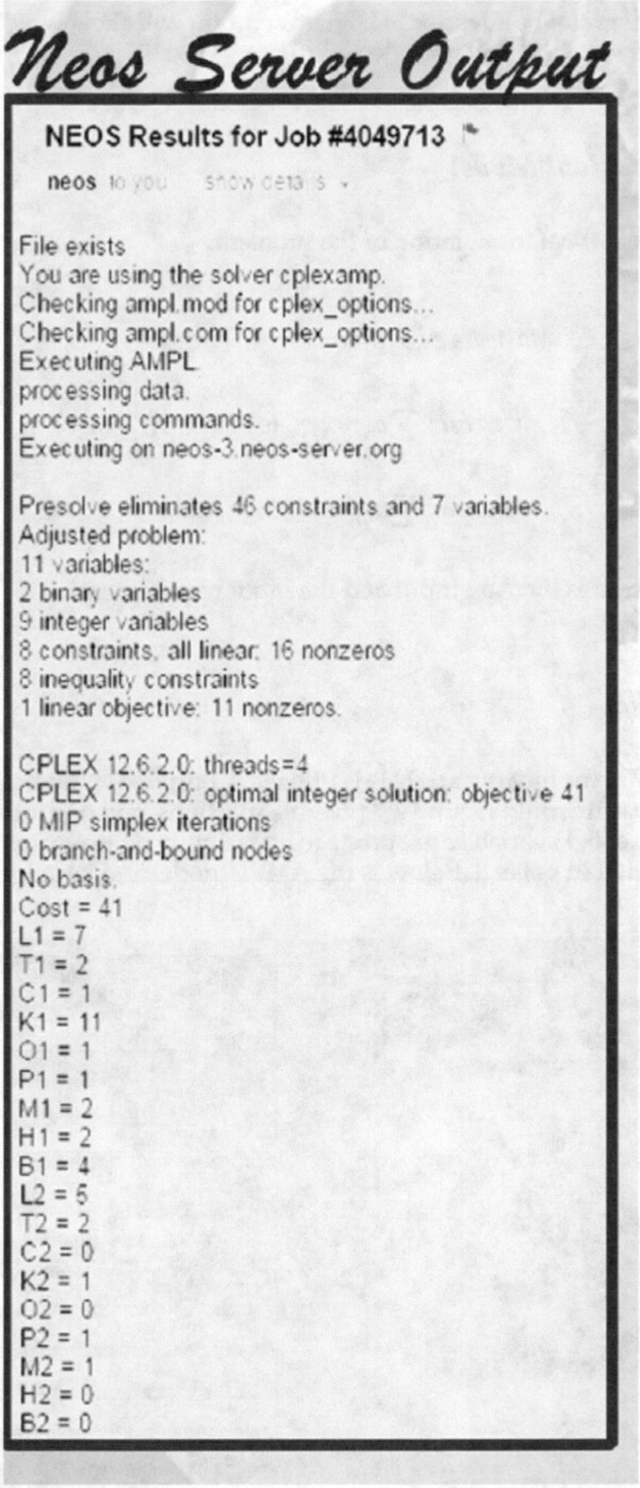

Neos Server Output

NEOS Results for Job #4049713

neos to you   show details

File exists
You are using the solver cplexamp.
Checking ampl.mod for cplex_options...
Checking ampl.com for cplex_options...
Executing AMPL
processing data.
processing commands.
Executing on neos-3.neos-server.org

Presolve eliminates 46 constraints and 7 variables.
Adjusted problem:
11 variables:
2 binary variables
9 integer variables
8 constraints, all linear; 16 nonzeros
8 inequality constraints
1 linear objective; 11 nonzeros.

CPLEX 12.6.2.0: threads=4
CPLEX 12.6.2.0: optimal integer solution: objective 41
0 MIP simplex iterations
0 branch-and-bound nodes
No basis.
Cost = 41
L1 = 7
T1 = 2
C1 = 1
K1 = 11
O1 = 1
P1 = 1
M1 = 2
H1 = 2
B1 = 4
L2 = 6
T2 = 2
C2 = 0
K2 = 1
O2 = 0
P2 = 1
M2 = 1
H2 = 0
B2 = 0

Fig. 4.24 Newman Inventory Output

one-day tournaments can be attended. The final solution will show which tournaments are selected (given the value (1) and not selected (0).

4.10.3 Mathematical Model

Below is the mathematical formulation of the problem.

$$Maximize \; z : \sum_{j}^{n} p_j * x_j \tag{4.35}$$

$$s.t. : \; cost : \sum_{j}^{n} c_{i,j} * x_j \leq b - i \, \forall i \in 1,\ldots,m \tag{4.36}$$

$$duration : \sum_{j}^{n} a_{i,j} * x_j \geq 5 \tag{4.37}$$

Figure 4.25 illustrates the App input and the input requirements.

4.10.4 Algorithm

The NEOS algorithm for binary variables is utilized. Complete enumeration is ruled out of the question because it would require 2^{10} possible solutions, and one must be careful in the constraints since the $(0,1)$ variables are prone to infeasibility if the resources are inadequate. But that is hard to tell in general. Below is the AMPL model and its solution:

Fig. 4.25 Tournament Selection App

4.10.5 Solution App

Figure 4.26 illustrates the blocks programming for the tournament scheduling. The text in Figure 4.27 illustrates the solution from NEOS.

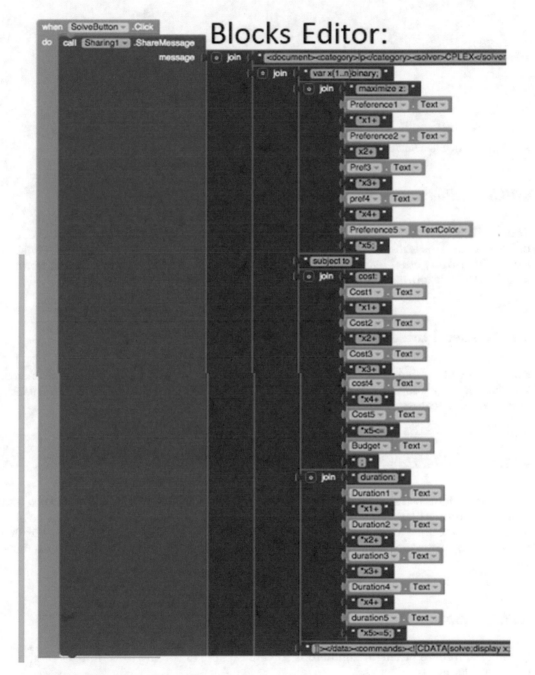

Fig. 4.26 Tournament Selection Blocks

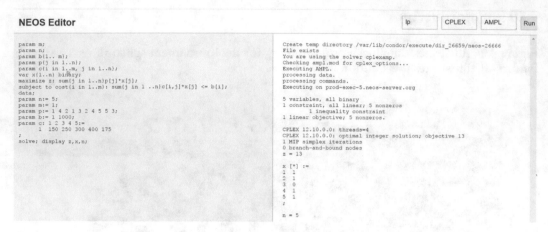

Fig. 4.27 Tournament Solution

4.10.6 Evaluation

The application works through the use of App Inventor2, AMPL, and the NEOS server out of Wisconsin. This is an effective method that can be used for not only the Ultimate Frisbee team, but also any other team administration that wishes to find a more precise and optimal solution to their tournament scheduling.

4.11 General Integer Programming Problems

As we showed in Chapter 3, we sometimes need integer solutions to our general problems. We show how this can be carried out with some simple changes in the AMPL file. This is a very desirable feature of AMPL.

4.11.1 Introduction

As we have demonstrated IP problems are challenging and will require either branch and bound or cutting plane techniques to solve the problem. Certain times we might be able to round up/round down the continuous variables to their integer counterparts, but this is always happenstance and may not guarantee an optimal solution. So, we must do more work.

4.11.2 Problem

In the NEOS server, there are numerous solvers capable of solving these integer programming problems.

4.11.3 Mathematical Model

Our LP mathematical model basically stays the same, except for the declaration of the integer variables.

$$Maximize \ \ Z = \sum_j c_j x_j \tag{4.38}$$

$$s.t. : \sum_j a_{ij} x_j \leq b_i \ \forall i \tag{4.39}$$

$$x_j \geq 0 \ \forall j \ and \ integer \tag{4.40}$$

For instance, these blocks shown in the Figure show we can treat the objective function and constraints and alter them to fit our model, please see Figure 4.28.

4.11.4 Algorithm

We must modify the AMPL problem file with a simple addition of the word *integer* aside each variable that needs to be integer along with the greater than or equal to ≥ 0 inequality.

The nonnegative inequality is important to prevent the algorithm from using certain integer variables that will create an unbounded solution. So, we must include this inequality in the declaration. If it is not included, we may get lucky with our constraints, but there is no guarantee that this will work (Fig. 4.29).

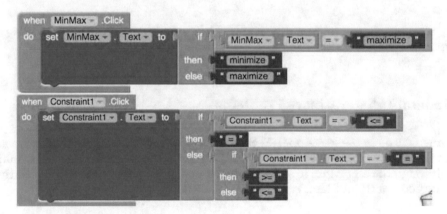

Fig. 4.28 Inequalities Flexibility

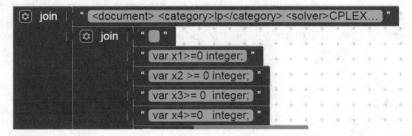

Fig. 4.29 Integer Inputs

4.11.5 Solution App

Let's solve the following problem presented at the end of Chapter 3 where we examined the solution of a general LP:

$$\text{Maximize } Z = 9x_1 + 12x_2 + 6x_3 + 4x_4 \tag{4.41}$$

$$s.t.: \; 5x_1 + 8x_2 + 6x_3 + 3x_4 \leq 68 \tag{4.42}$$

$$x_1 - 2x_2 + 3x_3 - 4x_4 \geq 26 \tag{4.43}$$

$$x_j \geq 0 \; \forall j \text{ and integer} \tag{4.44}$$

The optimal LP solution returned by the server is an upper bound on the problem:

$$Z = 89.33 : x_1 = 5.33, x_2 = 0, x_3 = 6.89, x_4 = 0 \tag{4.45}$$

When we solve this on the NEOS server, we get:

$$Z = 87 : x_1 = 5, x_2 = 0, x_3 = 7, x_4 = 0 \tag{4.46}$$

So this solution makes eminent sense and is very close to the LP upper bound. One can imagine the branch and bound algorithm first branching on the x_1 variable since it has the largest c_j generating a lower bound, then branching on the x_3 variable to finally find the best lower bound.

4.11.6 Evaluation

This is a great benefit to our App process for solving difficult problems. Now let's see how to incorporate binary variables.

4.12 General Integer Binary Problems

In the same way as in the last App in Section 4.11, to solve a problem with binary variables is similar to what we did with just the integer variables. The solution process is somewhat similar although the algorithm to solve the problem may be different or it could be the same as with the general IP problem.

4.12.1 Introduction

Binary variables $x_j \in (0,)$ are excellent modeling tools for many practical applications. Again, H. Taha's Chapter [14] on Integer Programming goes over many examples where binary variables allow us to model many different situations. One can be very creative with these variables.

4.12.2 Problem

The problem with the binary variables is that feasibility may be difficult to solve for the resource constraints, so one has to be careful in the formulation process of the constraints.

4.12.3 Mathematical Model

The ILP model again is basically the same except for the declaration of the binary variables.

$$Maximize \ Z = \sum_j c_j x_j \tag{4.47}$$

$$s.t. : \sum_j a_{ij} x_j \leq b_i \ \forall i \tag{4.48}$$

$$x_j \qquad j \in 0, 1 \forall j \tag{4.49}$$

Again, we may have a mix of integer, binary, and continuous variables, but that is all possible with AMPL simply by defining the variables in the ways we have shown.

4.12.4 Algorithm

Figure 4.30 illustrates what must be done with the AMPL email file to declare the binary variables. The nonnegativity constraint is not necessary.

4.12.5 Solution App

Let's examine a simple problem for solution:

$$Maximize \ Z = 9x_1 + 12x_2 + 6x_3 + 4x_4 \tag{4.50}$$

$$s.t. : \ 5x_1 + 8x_2 + 6x_3 + 3x_4 \leq 18 \tag{4.51}$$

$$2x_1 - 2x_2 + 3x_3 - 4x_4 \geq 4 \tag{4.52}$$

$$x_j, j \in 0, 1 \forall j \tag{4.53}$$

One can see that if you simply sum the variables of the first constraint, that the minimum value of the right-hand side is very sensitive to the $(0, 1)$ phenomenon. It is especially critical for constraint #2, but this is more difficult to see because of the negative signs.

If we solve the problem, we get

$$Z = 15 : x_1 = 1, x_2 = 0, x_3 = 1, x_4 = 0 \tag{4.54}$$

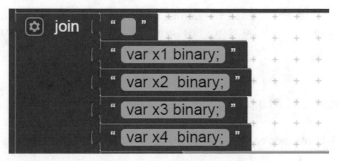

Fig. 4.30 Binary Inputs

4.12.6 Evaluation

The Binary feature of AMPL again is very powerful and simple to incorporate. We will now examine a most difficult IP with a quadratic objective function and zero-one variables.

4.13 Quadratic Assignment Problem

The penultimate problem in the IP part concerns the Quadratic Assignment Problem (QAP) which is a challenging nonlinear assignment problem in (0,1) variables. It is therefore an integer nonlinear programming problem which to say the least is $\mathcal{NP}$-Hard. It is fundamental to most layout problems. It is a combination of integer and nonlinear programming problems. If the number of facilities is small, then enumeration is one feasible way to solve the problem, whereas heuristics are necessary for large problems. We shall examine heuristics later on in the book.

This problem is also something that a standard LP or NLP approach with AMPL will not be feasible, so a novel approach is needed. It is a notoriously difficult problem, but we will look at small instances and use AI2 to come up with an approach for its solution.

4.13.1 Introduction

It is insightful to look carefully at this problem since it reveals much. If we have N=4 facilities, we have a given distance and flow matrix. The distance matrix as defined by the possible configuration is as shown in Figure 4.31. We will assume that the distance and flow matrices are symmetric which is not unreasonable for most problems.

$$
\mathbf{D} = \begin{array}{c} \\ 1 \\ 2 \\ 3 \\ 4 \end{array}
\begin{array}{cccc}
1 & 2 & 3 & 4 \\
\left[\begin{array}{cccc}
0 & 80 & 150 & 170 \\
80 & 0 & 130 & 100 \\
150 & 130 & 0 & 120 \\
170 & 100 & 120 & 0
\end{array}\right]
\end{array}
\qquad
\mathbf{F} = \begin{array}{c} \\ 1 \\ 2 \\ 3 \\ 4 \end{array}
\begin{array}{cccc}
1 & 2 & 3 & 4 \\
\left[\begin{array}{cccc}
0 & 5 & 2 & 7 \\
5 & 0 & 3 & 8 \\
2 & 3 & 0 & 3 \\
7 & 8 & 3 & 0
\end{array}\right]
\end{array}
$$

4.13.2 Problem

We wish to arrange the activities in the layout to minimize the overall transportation flow cost. There are $n!$ possible solutions or permutations, so the number of solutions skyrockets with large N.

4.13.3 Mathematical Model

We need to use $(0, 1)$ integer variables along with a quadratic objective function. Our objective function is

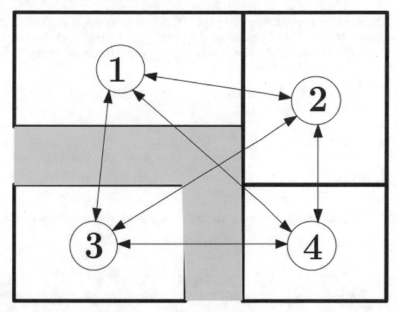

Fig. 4.31 QAP N=4

$$Z = \sum_i \sum_j \sum_k \sum_\ell f_{ik} d_{j\ell} x_{ik} x_{j\ell} \tag{4.55}$$

$$s.t. \quad \sum x_{ij} = 1 \forall i \tag{4.56}$$

$$\sum x_{ij} = 1 \forall j \tag{4.57}$$

$$x_{ij} \in (0,1) \forall i,j \tag{4.58}$$

While it appears to have only the simple assignment constraints, the number of possible solutions is complicated by the objective function.

4.13.4 Algorithm

Many researchers have tried valiantly to crack the problem with sophisticated branch and bound approaches. We shall not develop a sophisticated solution, but use backtracking and enumeration as we did in the TSP problem since it is nice to have the optimal solution. One can use the objective function to iterate all the solutions in AI2 for the $N! = 4! = 24$ solutions and that is in essence what we have done.

4.13.5 Demonstration

Figure 4.32 illustrates part of the blocks used to capture the objective function minimum for the QAP assignment problem. While the objective function values of the 4! solutions are computed with the math blocks and stored in a list, we simply have to find the minimum in the list along with its index. For the example problem described above with the distance matrix **D** and flow matrix **F**, the program yielded the optimal topology of $\mathbf{x} = 4,1,3,2$ with objective value of $Z = 3,260$. It computed the solution very quickly.

Figure 4.33 illustrates the input screen with the distance and flow matrices. Below the matrices are the Winner, Objective, and Permute buttons. After inputting the distance and flow matrix values in the upper triangular part of the screen, one pushes the Objective button and it will compute the backtrack values of the objective function and then one pushes the Permute button to get the optimal answer and the objective function value. In this case, for the input data, the optimal layout is

$$1 \Rightarrow 4 \Rightarrow 3 \Rightarrow 2, \qquad Z = 3,260$$

see Figure 4.33.

4.13.6 Evaluation

The AI2 program works pretty well even though it is essentially an enumeration/backtrack algorithm.

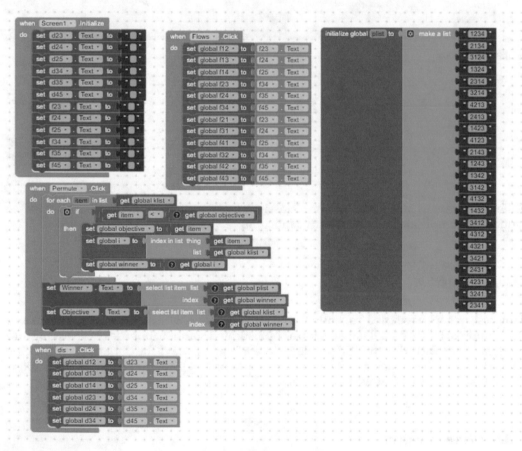

Fig. 4.32 QAP App with Procedure for Finding the Minimum Layout

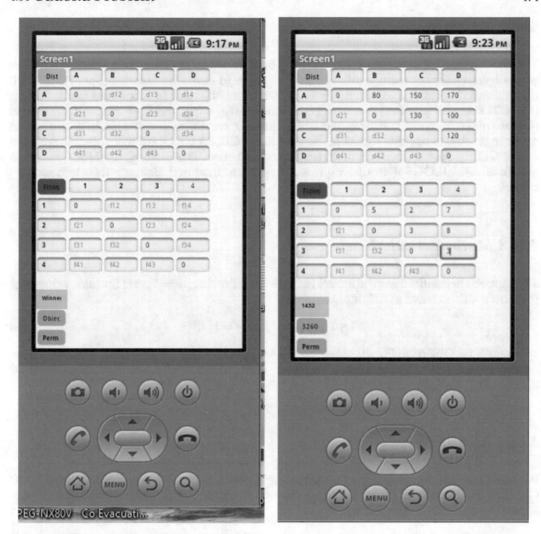

Fig. 4.33 QAP Example Input App and Solution

4.14 Sudoku Problem

For our final problem, we will examine a Sudoku puzzle problem. Sudoku while not strictly an optimization problem as there is no objective function, still requires an integer programming formulation and algorithm for its solution so it is well placed in this Chapter.

Italo DeSouza created this App in 2016 for the MIE 379 class. He is responsible for finding the link between NEOS and AI2.

4.14.1 Introduction

In the hope of inspiring and challenging students to go beyond the traditional textbook problems, the ultimate model chosen was a Sudoku Solver. While it is a more challenging model to build, it certainly is more entertaining and fun. The Sudoku solver can solve any Sudoku puzzle from the easiest to the most difficult. The user inputs the preset Sudoku values from the puzzle they want to solve and leaves the unknown values as 0 or blank

4.14.2 Problem

This Sudoku model is different than the previous two models in two major ways. First, there is no objective function since all the model needs is to find a feasible solution. AMPL will understand this command simply by not entering an objective function and just leaving it blank. The second way that this model is different is that it requires the function *alldifferent* in order to make the rows, columns, and Latin squares have different values from each other. In order to be able to use this function, the IBM ILOG CPLEX CP Optimizer solver is required as the CPLEX solver does not take this type of function. Fortunately, the NEOS server is able to take the ILOG solver making this App work as intended. Below is the mathematical model.

4.14.3 Mathematical Model

The input data required for the model include the following: m_j = Preset Sudoku values (for unknown values leave as "." or 0) Variables:

$$1 \leq x_{ij} \leq 9 \, i = 1, .., 9; j = 1, .., 9 \tag{4.59}$$

There is no Objective Function per se. We are only looking for a feasible solution. Thus:
Subject to: Fill in the hints: $\forall m_{ij} > 0 x_{ij} = m_{ij}$ Latin Square: $\forall \ i different \ \forall \ j x_{ij}$ Regions: $\forall (i \in 0, 1, 2, \forall \ j0, 1, 2)$ *all different* $r \in i * 3 + 1 \ldots 3 + 3, 3 + 1 \ldots c \in j * 3 + 3) * x_{rc}$

4.14.4 Algorithm

We must use the mixed-integer programming algorithm to solve the problem. IBM ILOG CPLEX CP Optimizer: ILOG CP Options: optimizer=cp all diffinferencelevel=4 debug-expr=0 logperiod=1 logverbosity=0

4.14.4.1 Demonstration

Below are the parameters and the mathematical model in AMPL. This AMPL model was developed from NEOS.

```
AMPL Model
param m >= 1, integer, default 3;
param n := m*m;
set N := 1..n;
param P{N,N} default 0, integer, >= 0, <= n;
var z{N,N,N} binary;
minimize obj: 0;
subject to col_sum{j in N, k in N}:
    sum{i in N} z[i,j,k] = 1;
subject to row_sum{i in N, k in N}:
    sum{j in N} z[i,j,k] = 1;
subject to sqr_sum{r in 0..m-1, c in 0..m-1, k in N}:
   sum{p in 1..m, q in 1..m} z[m*r+p,m*c+q,k] = 1;
subject to unique{i in N, j in N}: sum{k in N} z[i,j,k] = 1;
subject to fixed{i in N, j in N: P[i,j] <> 0}:
    z[i,j,P[i,j]] = 1;
data:
param m:= 3;
param P:  1   2   3 4   5   6 7   8   9 :=
```

```
1           1  .  .  .  .  6  3  .  8
2           .  .  2  3  .  .  .  9  .
3           .  .  .  .  .  .  7  1  6

4           7  .  8  9  4  .  .  .  2
5           .  .  4  .  .  .  9  .  .
6           9  .  .  .  2  5  1  .  4

7           6  2  9  .  .  .  .  .  .
8           .  4  .  .  .  7  6  .  .
9           5  .  7  6  .  .  .  .  3 ;
commands:
solve;
#display the results
for {i in N}{
    for {j in N}{
        for {k in N}{
            if (z[i,j,k] == 1) then printf "%3i", k;
        };
        if ((j mod m) == 0) then printf " | ";
    };
    printf "\n";
    if ((i mod m) == 0) then {
        for {j in 1..m}{
            for {k in 1..m-1}{ printf "---" };
            if (j < m) then
                printf "----+-";
            else
                printf "----+\n";
        };
    };
};
```

Figure 4.34 illustrates the input screen while Figure 4.35 illustrates the programming blocks. Figure 4.36 illustrates the solution.

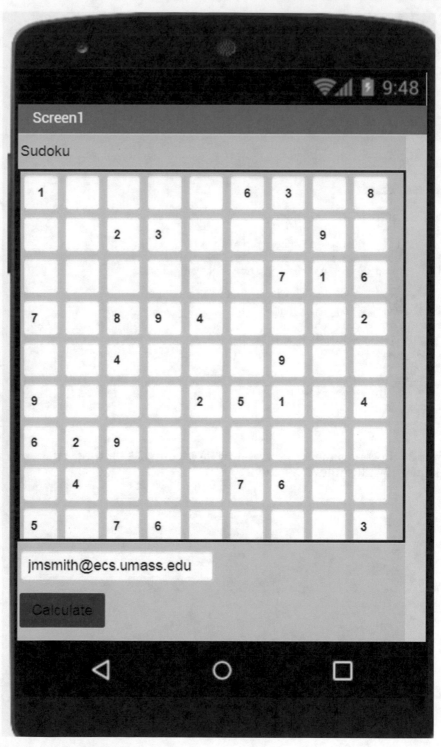

Fig. 4.34 Input of Sudoku Problem

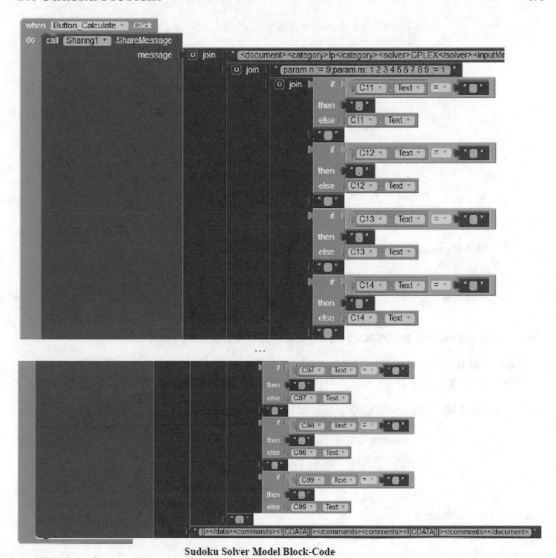

Sudoku Solver Model Block-Code

Fig. 4.35 Blocks of Sudoku Problem

4.14.5 Evaluation

While it takes the fun out of solving the Sudoku problem by hand, for a very complex MIP, the App does an excellent job and allows the user to check her/his solution to the problem.

```
Solution determined by presolve;
objective = 0.
  1  7  5 |   4  9  6 |   3  2  8 |
  8  6  2 |   3  7  1 |   4  9  5 |
  4  9  3 |   8  5  2 |   7  1  6 |
---------+-----------+-----------+
  7  1  8 |   9  4  3 |   5  6  2 |
  2  5  4 |   1  6  8 |   9  3  7 |
  9  3  6 |   7  2  5 |   1  8  4 |
---------+-----------+-----------+
  6  2  9 |   5  3  4 |   8  7  1 |
  3  4  1 |   2  8  7 |   6  5  9 |
  5  8  7 |   6  1  9 |   2  4  3 |
---------+-----------+-----------+
```

Fig. 4.36 Sudoku Solution

4.15 Exercises

We will examine a number of Integer Programming problems.

1. **Knapsack Problem:**
   Here, we require that the knapsack have integer amounts.

   a) Solve the following (knapsack) problem:

   $$\text{Maximize } 2x_1 + 3x_2 + 8x_3 + x_4 + 5x_5 \tag{4.60}$$
   $$s.t. : 3x_1 + 7x_2 + 12x_3 + 2x_4 + 7x_5 \leq 10 \tag{4.61}$$
   $$x_1, x_2, x_3, x_4, x_5 \geq 0 \text{ and integer} \tag{4.62}$$

2. **Binary Problem**

   $$\text{Minimize} f(\bar{x}) = 3x_1 + 2x_2 + 7x_3 \tag{4.63}$$
   $$s.t. : 1x_1 + x_2 + x_3 \geq 6 \tag{4.64}$$
   $$3x_1 - x_2 - x_3 \geq 1 \tag{4.65}$$
   $$x_1, x_2, x_3 \in 0, 1 \tag{4.66}$$

3. **LP and IP Problem (after Fourer et.al. [7])**

   $$\text{Maximize} f(\bar{x}) = 5x_1 + 8x_2 \tag{4.67}$$
   $$s.t. : 1x_1 + x_2 \leq 6 \tag{4.68}$$
   $$5x_1 + 9x_2 \leq 45 \tag{4.69}$$
   $$x_1, x_2 \geq 0 \tag{4.70}$$

First, graphically solve the problem for the optimal solution, then solve it with Linear Programming. Finally, solve it where x_1, x_2 are integer variables.

4. Fixed Charge Problem

A Fixed Charge problem is a mixed-integer programming problem with the following formulation:

$$\text{Minimize } Z = \sum_j f_j(x_j) = \sum_j (k_j + c_j x_j) \tag{4.71}$$

$$\text{subject to}: \sum_j a_{ij} \leq b_i \ \forall i \tag{4.72}$$

$$x_j \geq 0 \tag{4.73}$$

$$\text{and } y_j = \begin{cases} 0 & \text{if } x_j = 0 \\ 1 & n = 0 \end{cases} \tag{4.74}$$

Let's suppose we have three phone service providers that can provide service to our area: ATT, Verizon, and T-Mobile.

- ATT (#1) charges a fixed price of $16/month plus $0.25 per minute.
- Verizon (#2) charges a fixed price of $25/month and $0.21/minute.
- T-Mobile (#3) charges a fixed price of $18/month and $.22/minute.

Let $y_j = \begin{cases} 1 & \text{if company j is selected} \\ 0 & otherwise \end{cases}$

Let's define x_j as the amount of service time

$$x_j \text{ number of call minutes with provider } j$$

So for the service time we have the following equation:

$$x_1 + x_2 + x_3 = 200 \tag{4.75}$$

The maximum subscriber time x_j is 200 minutes for provider j only if the provider is selected for service:

$$0 \leq x_j \leq 200y_j \ \forall \ j \tag{4.76}$$

Formulate this example problem and solve it with the given appropriate App.

5. **Manufacturing Processes:**

Take the manufacturing processes App and add a cost constraint on the problem and re-solve for the optimal integer values.

6. **Food Produce Inventory Problem:**

Like the model at the Newman Center, find a restaurant that would allow you to model their ordering processes for producing items similar to the App described in this Chapter.

7. **Mobile Food Truck Scheduling:**

Find the mobile food trucks or processes used on your campus for delivering meals to the students. Based on the number of stops and the estimated location of students at the stops, determine the sequence of locations this mobile process should be used to deliver meals to the students in the most efficient way possible, starting from the first stop to the last.

8. **Quadratic Assignment Problem:**
   Utilizing the Quadratic Assignment App, create a couple of examples with $N = 4$ along with a Flow Matrix and Distance matrix of your choice and find their optimal solutions.

9. **Quadratic Assignment Problem For $N = 5$ facilities in the QAP:** Develop an App for the QAP, $N = 5$. Generate a sample layout for the allocation. How many explicit solutions must you examine in your App for complete enumeration, with $N = 5$? You should follow a similar methodology as in the development of the App for $N = 4$ as described in the text.

Nonlinear Programming
$$\sum_i \sum_j c_{ij} x_i x_j, \ \forall \ (i,j)$$

Overview Nonlinear Programming (NLP) problems represent some of the most complex optimization problems. Especially when contrasted with LP problems which have dominated the literature in Operations Research. This is because some of the nice properties such as monotonicity and convexity in Linear Programming are not always possible with NLP problems, so many local optimal solutions will exist, not just a single global optimization one. Figure 5.1 illustrates an example problem and the usual local vs. global phenomenon.

Minimize $(Z = x_1 * x_4 * (x_1 + x_2 + x_3) + x_3, x_4^2 + x_1^2 + x_2^2 + x_3^2 = 40, 25 <= x_1 * x_2 * x_3 * x_4,$
Solution $Z = 16, X_1 = 1, x_2 = 5, x_3 = 5, x_4 = 1$ (green dot in right hand corner of left figure))

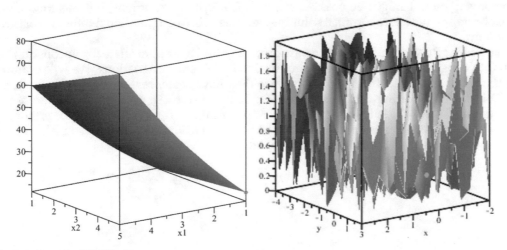

Fig. 5.1 Complex NLP Problem

Keywords: Algorithms, Unconstrained, Constrained

© Springer Nature Switzerland AG 2021
J. MacGregor Smith, *Combinatorial, Linear, Integer and Nonlinear Optimization Apps*,
Springer Optimization and Its Applications 175,
https://doi.org/10.1007/978-3-030-75801-1_5

Since the fabric of the universe is most perfect, and is the work of a most wise Creator, nothing whatsoever takes place in the universe in which some form of maximum and minimum does not appear.

—LEONARD EULER

The great watershed in optimization isn't between linearity and nonlinearity, but convexity and nonconvexity

—R.T. ROCKAFELLAR

Baseball players or cricketers do not need to be able to solve explicitly the nonlinear differential equations which govern the flight of the ball. They just catch it.

—PAUL ORMEROD

5.1 Introduction

Nonlinear programming problems tend to be some of the most challenging problems we face in optimization. The textbook by Hamdy Taha discusses the general nature of NLP problems [14]. There are many, many applications for NLP type problems. In fact, LP is a special case of NLP. The general structure is the following programming problem.

$$\text{Max or Min } f(\mathbf{x}) \tag{5.1}$$
$$s.t. : \text{inequalities } g_i(\mathbf{x}) \leq 0 \tag{5.2}$$
$$\text{equalities } h_i(\mathbf{x}) = 0 \tag{5.3}$$
$$\text{nonnegativity } \mathbf{x} \geq 0 \tag{5.4}$$

The form of these functions can be quite complex. If the objective functions and constraints are convex, then it might be possible to get global optimal solutions, but in general, one must be happy with local optimal solutions because of the computational difficulty of these problems.

Nonlinear Programming methods are quite vast and are appropriate for unconstrained, equality constrained, and inequality constrained problem formulations. We will examine a number of these nonlinear problems since they have many real-world applications, especially for engineering design applications. In some of these example Apps, the use of calculus affords us directly the equations for optimization. The equations can be nonlinear, but if the expressions are not too complex, AI2 can be utilized. If the problem is complex, then the NEOS server approach is recommended.

5.2 Group Testing for COVID-19

Group Testing (GT) is often used for conducting tests on viruses when the cost of testing is high and the probability of infection is low. During COVID-19 pandemic researchers from the Technion [3] proposed a simple accelerated scheme to speed up COVID-19 testing. COVID-19 is diagnosed with polymerase chain reaction (PCR) testing, which is common for virus monitoring.

5.2.1 Introduction

This test examines the presence of a unique genetic sequence of viruses in a sample taken from the patient. The test takes several hours thus generating a bottleneck in identifying COVID-19 infected people. The standard method of testing is to examine every sample taken from every individual.

In the new pooling approach that the Technion researchers tested in February 2020, each sample is first divided—one part, if needed, is kept in reserve for later and the other part is commingled with a number of other samples; the molecular testing is then performed on an aggregated entity. If this test is negative all individuals in the aggregate are negative. Else, when the joint sample is found to be positive, they conduct a test (using the part kept in reserve) for each of the individuals included in the aggregate. See Figure 5.2.

According to Dr. Yuval Gefen, director of the Rambam Clinical Microbiology Laboratory, [1] *Today, we receive approximately 200 COVID-19 test samples a day, and each sample undergoes individual examination. According to the new pooling approach that we have currently tested, molecular testing can be performed on a combined sample, taken from 32 or 64 patients. This way we can significantly accelerate the testing rate. Only in those rare cases, where the joint sample is found to be positive, will we conduct an individual test for each of the specific samples.*

5.2.2 Problem

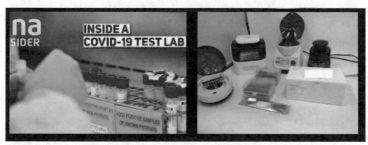

Fig. 5.2 Group Testing after [6]

The goal of the GT is to identify (and then quarantine) all the individuals infected. The sample is taken from all but the test is in batches or pools (the tests are time consuming and normally expensive). So it make sense, that if the probability of infection p is very small, and if the samples can be lab combined (somehow mixed together without losing same accuracy), then GT will be beneficial. Figure 5.3 illustrates the input screen with all the parameters. One must enter: c, K, p, and n and then submit them to the blocks (Figure 5.4).

---

[1] https://www.technion.ac.il/en/2020/03/pooling-method-for-accelerated-testing-of-covid-19/

Fig. 5.3 Input Screen

5.2.3 Mathematical Model

If you want to test entire population for COVID-19 and there is only a small number of individuals already infected. GT could lead to a substantial savings in time and money. The mathematical model is based upon the binomial distribution and is related to binary search.

Assume : $p = P(test\,positive)$ where
n := number of tests combined for a group testing
Then:

$$P(group\ tests\ positive) = P(at\ least\ one\ positive\ individual\ in\ the\ group) \tag{5.5}$$

$$= 1 - P(no\ positive\ individuals\ in\ the\ group\ of\ n) \tag{5.6}$$

$$= 1 - P(all\ individuals\ in\ the\ group\ are\ negative \rightarrow) \tag{5.7}$$

$$= 1 - (1-p)^n \tag{5.8}$$

$$L = Expected\#\ of\ tests\ needed\ to\ identify\ positives\ in\ the\ group\ of\ n\ individuals \tag{5.9}$$

$$= 1 * (1-p)n + (n+1) * (1-(1-p)n) = 1 + n * (1-(1-p)^n) \tag{5.10}$$

The cost contributed to the current standard (process all samples) for n individuals is $n := c * n$. Under the new proposed accelerated testing scenario, the cost (for the group of n) will be

$$c * L(n) + K \tag{5.11}$$

where K denotes a small fixed cost contributing to some extra work (*e.g.*, mixing the samples, administration).

Suggested objective might be

$$Minimize\ Z = c * n - L(n) - K \tag{5.12}$$

and if we take

$$c = 1 \text{ and } K = 0 \tag{5.13}$$

we can minimize the percentage of average number of tests saved by the group test:

$$Minimize\ Z = 1 * (n - L(n)) \tag{5.14}$$

5.2.4 Algorithm

The App calculates the Expected # of positive tests for a given population of n patients from the equations above and returns the expected savings with GT. Here are the programming blocks in Figure 5.4:

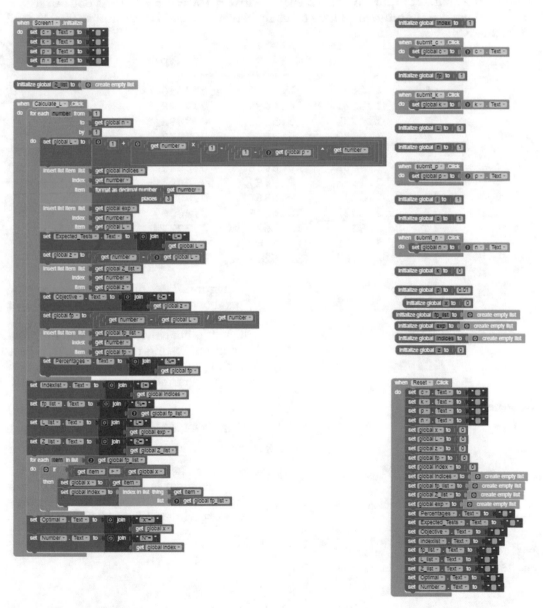

Fig. 5.4 Blocks Programming

5.2.5 Solution App

An important characteristic of GT is that the probability of getting infected should be low, otherwise, GT may not be effective and complete testing of the population is necessary. So, for example, if p is very small say $p = 0.01 (1\%)$ and $n = 16$ then $L = 3.37$ meaning that we save about 12 tests per group of 16 (i.e. an 80% savings!) Here, we construct a do loop for carrying out our calculations:

For i from 5 by 1 to 16 do

$$L[i] := 1 + i * (1 - (1-p)^{(i)}) : \text{(expected number of tests)} \tag{5.15}$$

$$f[i] := c*i - c*L[i] - K : \text{(objective function)} \tag{5.16}$$

$$fp[i] := 1 - ((1 + i(1 - p^{(i)}))/(i)) : \text{(percentage savings)} \tag{5.17}$$

Below is the tabular output from this scenario with $c = 1, K = 0, p = 0.01$ As one can see we gain an 80 per cent improvement here by using between 10 *and* 11 tests rather than the full 16 (Figure 5.5).

i	L	f	f(\%p)	
5	1.2450	3.7550	0.7510	
6	1.3511	4.6489	0.7748	
7	1.4755	5.5245	0.7892	
8	1.6180	6.3820	0.7977	
9	1.7783	7.2217	0.8024	
===>10	1.9562	8.0438	0.8044<===	
===>11	2.1513	8.8487	0.8044<===	11 is optimal
12	2.3634	9.6366	0.8031	
13	2.5922	10.4078	0.8006	
14	2.8376	11.1624	0.7973	
15	3.0991	11.9009	0.7934	
16	3.3767	12.6233	0.7890	

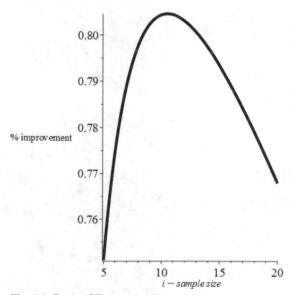

Figure 5.5 nicely illustrates the parabolic curve for the optimal number of tests which numerically is $10.516 \approx 11$. On the other hand, let's say that the probability of testing positive is 0.02, a little larger probability, then for the other given parameters and the range of from $1 to 16$:

$$c = 1, K = 0, p = 0.02, n = 16 \tag{5.18}$$

$$L = 5.42, Z = 10.58, F(\%) = 0.66\%, N^* = 8 \tag{5.19}$$

which makes sense that the percentage improvement is around 73% which is a little less when compared with the previous experiment (Figure 5.6).

Fig. 5.5 Optimal Percentage Improvement

i	L	f	fp
1	1.0200	-0.0200	-0.0200
2	1.0792	0.9208	0.4604
3	1.1764	1.8236	0.6079
4	1.3105	2.6895	0.6724
5	1.4804	3.5196	0.7039
6	1.6849	4.3151	0.7192
7	1.9231	5.0769	0.7253
8	2.1939	5.8061	0.7258 <===== 8 is optimal
9	2.4963	6.5037	0.7226
10	2.8293	7.1707	0.7171
11	3.1920	7.8080	0.7098
12	3.5834	8.4166	0.7014
13	4.0027	8.9973	0.6921
14	4.4490	9.5510	0.6822
15	4.9215	10.0785	0.6719
16	5.4192	10.5808	0.6613

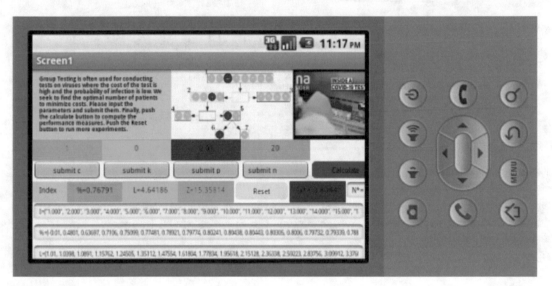

Fig. 5.6 Output Screen

The Hessian of the optimal patient function is given below which demonstrates that the function is a continuous monotonically increasing function of the variables n, p. Figure 5.7 demonstrates the smooth curvature of the optimal patient function.

$$
\mathbf{H} = \begin{bmatrix} -\dfrac{-2\,p^n \ln(p) - n\,p^n (\ln(p))^2}{n} + 2\,\dfrac{1 - p^n - n\,p^n \ln(p)}{n^2} - 2\,\dfrac{1 + n(1 - p^n)}{n^3} & -\dfrac{1}{n}\left(-2\,\dfrac{n\,p^n}{p} - \dfrac{p^n n^2 \ln(p)}{p}\right) - \dfrac{p^n}{p} \\[2ex] -\dfrac{1}{n}\left(-2\,\dfrac{n\,p^n}{p} - \dfrac{p^n n^2 \ln(p)}{p}\right) - \dfrac{p^n}{p} & \dfrac{p^n n^2}{p^2} - \dfrac{n\,p^n}{p^2} \end{bmatrix}
$$

$$(5.20)$$

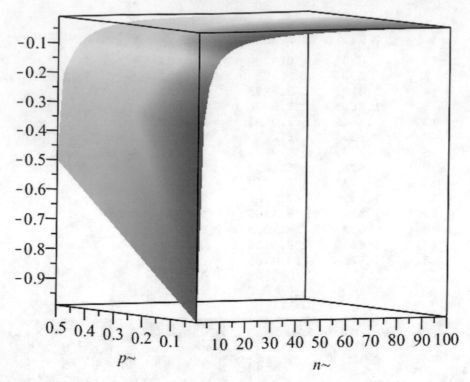

Fig. 5.7 Curvature of Optimal Patient Numbers Function

5.2.6 *Evaluation*

GT has limitations since if most pools test positive, it will lead to a number of individual tests for the entire population, which defeats the purpose of GT. In general, however, GT can be very beneficial.

5.3 Gas Guzzler Nonlinear Program

This is a problem of trying to find the optimal speed with which to drive a car to minimize the cost of driving a car in relation to the maintenance and operation of the car. We shall use calculus to solve this unconstrained optimization problem.

5.3.1 Introduction

Driving and minimizing the operating costs of a car together with the cost of gasoline is an important task most people cherish because of the unpredictable fluctuation of gas prices. The idea for this App problem first appeared in a textbook on optimization by David Russel [12].

5.3.2 Problem

Let's say we wish to plan a long-distance trip and we wish to determine the optimal speed decision variable s with which we should drive during the duration of the trip, which has a given distance or contextual variable d.

We are concerned about our gas mileage because it has become so expensive to drive in the last few years. We want to first minimize $f_1(s)$ the costs of operating our car (rental car costs) and second minimize the mileage costs $f_2(s)$ and optimize our speed decision variable s to offset the rising cost of gasoline. There is a subtle tradeoff between the hourly operating costs of the car and the cost of gasoline. Thus, the total cost of our trip is

$$f(\text{total cost}) = f_1(s)(\text{operating \$}) + f_2(s)(\text{mileage \$})$$

There are no explicit constraints in the problem which makes it an unconstrained optimization problem. Unconstrained optimization problems are easier to solve than constrained problems.

One can show empirically that mileage per gallon m has the following relationship to speed s

$$m - v - \frac{s}{c} \tag{5.21}$$

- where $m :=$ mileage per gallon
- $v :=$ y-intercept of mileage speed cure
- $c :=$ Slope factor in the mileage speed curve.
- $h :=$ Operating costs of the car (\$)
- $g :=$ Cost of gas per gallon (\$)

From a study of the Federal Government which shows the gas mileage for different car-types on the y-axis and the speed of the vehicle on the x-axis. v in our model is the linear equation y-intercept in Figure 5.8. Thus, a linear approximation to the nonlinear gas/mileage function is appropriate.

5.3.3 Mathematical Model

- Further, from our operating cost and mileage cost expressions:

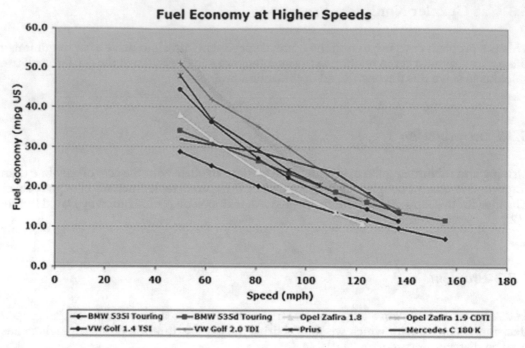

Fig. 5.8 Car Fuel Economy

$$f_1 = \frac{hd}{s} \rightarrow \frac{\$/\text{hr} * \text{miles}}{\text{miles/hr}} \tag{5.22}$$

$$f_2 = \frac{gd}{m} = \frac{gd}{v - \frac{s}{c}} \rightarrow \frac{\$/\text{gallon} * \text{miles}}{\text{miles/gallon}} \tag{5.23}$$

$$\text{total cost} = \frac{hd}{s} + \frac{gd}{v - \frac{s}{c}} \tag{5.24}$$

- If we differentiate the total cost function f and set to 0, we get

$$f'(s) = -\frac{hd}{s^2} + \frac{gd}{(v - \frac{s}{c})^2 c} = 0 \tag{5.25}$$

$$(a_1, a_2) \rightarrow a_1 = \frac{(-2h + 2\sqrt{hgc})cv}{2(-h + gc)} \tag{5.26}$$

$$a_2 = \frac{(-2h - 2\sqrt{hgc})cv}{2(-h + gc)} \tag{5.27}$$

- This is a quadratic equation, we can solve for s with the quadratic formula, since it will have two roots:

5.3.4 Algorithm

The AI2 blocks for the Gas Guzzler model are depicted in Figure 5.9. Again, because of the calculus equations it is pretty straightforward to program.

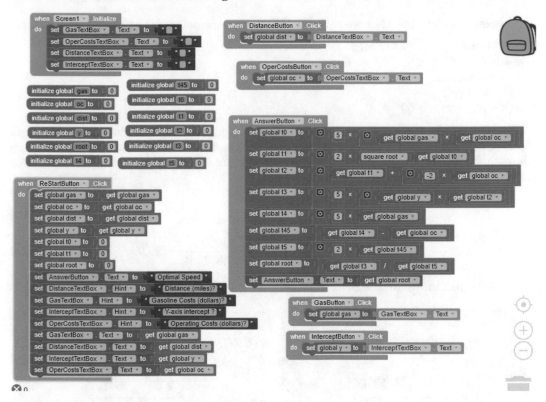

Fig. 5.9 Gas Guzzler Blocks

5.3.5 Demonstration

Figure 5.11 on the left illustrates the designer screen of the Gas Guzzler App. Figure 5.10 illustrates the actual convex objective function curve for our example problem.

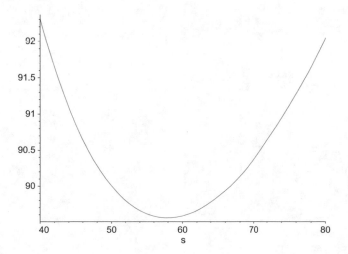

Fig. 5.10 Gas Guzzler App

For example, let's say that we have the following parameters for our example problem:

- $h = \$1$ operating costs/hour;
- $d = 1000$ miles;
- $c := 5$ a constant;
- $g := \$3.5/gal$;
- v (y − intercept) := 60

Then solving for s, we get two real roots: 57.878, −94.242 Obviously, the first positive root is our solution as is indicated in Figure 5.11. The convexity of the objective function is an interesting property of the problem as shown in Figure 5.11. If I plug in the parameters for the problem into the objective function

$$f \text{ totalcost} = \frac{hd}{s} + \frac{gd}{v - \frac{s}{c}} \tag{5.28}$$

$$f = \frac{1 * 1000}{57.878} + \frac{3.5 * 1000}{(60 - (57.878/5))} = 89.99 \tag{5.29}$$

which is the minimum total cost as indicated in Figure 5.11.

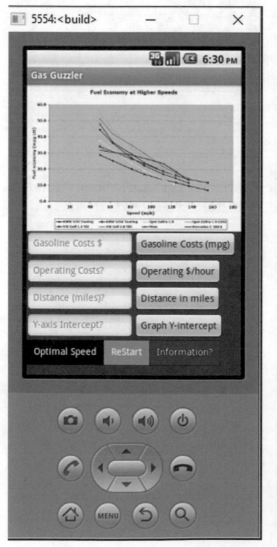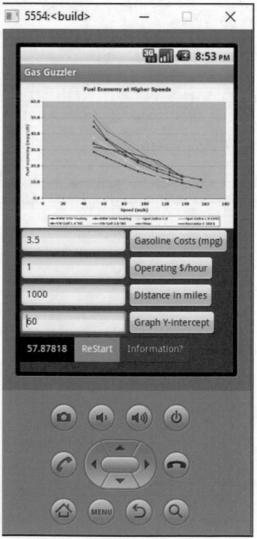

Fig. 5.11 Optimal Mileage Estimate

With the App one can resolve the problem for different intercept values v to see how sensitive the solution is for different cars used in travel.

5.3.6 Evaluation

This is a nice simple App with a straightforward solution. Of course coming up with the non-linear equations was the real trick. It is a useful App for an individual and quite appropriate for the phone environment.

5.4 Price is Right App

This is an interesting problem from Microeconomics, and we shall also use calculus to solve this problem. We have a demand curve and we seek to find the optimal price of an item subject to its demand and unit cost. This App shows the nice relationship between linear quantities and nonlinear quadratic functions.

5.4.1 Introduction

To demonstrate a simple model using the input parameters quantities price, demand, and unit cost and their interrelationships.

5.4.2 Problem

A firm's production department has found that the unit cost of its main product is c dollars. Meanwhile, the marketing department has estimated the relationship between the slope (p) and the maximum demand for the product maximum demand (D) (sales volume) follows the linear "demand curve", so that y the demand is a linear function of the price x. We need the following notation:

Variables	Description
$c :=$	Unit cost
$D :=$	Maximum Demand
$p :=$	Slope of demand curve
$x :=$	Selling price per unit
$y :=$	Unit demand
$z :=$	Overall objective function value

The demand is a linear function of the price with D intercept and slope p.

$$y = D - px \tag{5.30}$$

Figure 5.12 illustrates the linear relationship of price and demand.

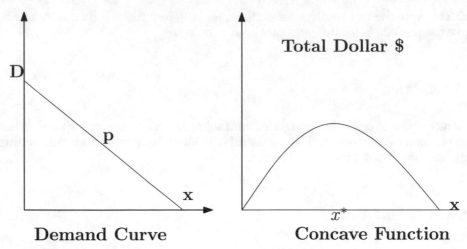

Demand Curve   Concave Function

Fig. 5.12 Price Demand Curve Relationships

5.4.3 Mathematical Model

The decision variable x is the selling price/unit in dollars. We would like to maximize our profit for the firm where we have the nonlinear relationship given by the following equation.

$$z = (x - c)y \tag{5.31}$$

Substituting the previous Equation for price and demand D for y and multiplying through with the above equation, we get

$$z = (x - c)(D - px) \tag{5.32}$$

or re-writing this we get a quadratic function the following Equation:

$$z = -px^2 + xD - Dc + pcx \tag{5.33}$$

If we differentiate this expression with respect to x and set it equal to 0, we get

$$z' = -2px + D + cp = 0 \tag{5.34}$$

and solving for the optimal price x, we get

$$x^* = \frac{D + cp}{2p} \tag{5.35}$$

The previous Equation represents the optimal price of the unit. This latter equation is the key to the entire App.

5.4.4 Algorithm

Figure 5.13 shows the complete set of blocks that are needed for this application. Because this is not a very complex algorithm, it is straightforward to program it.

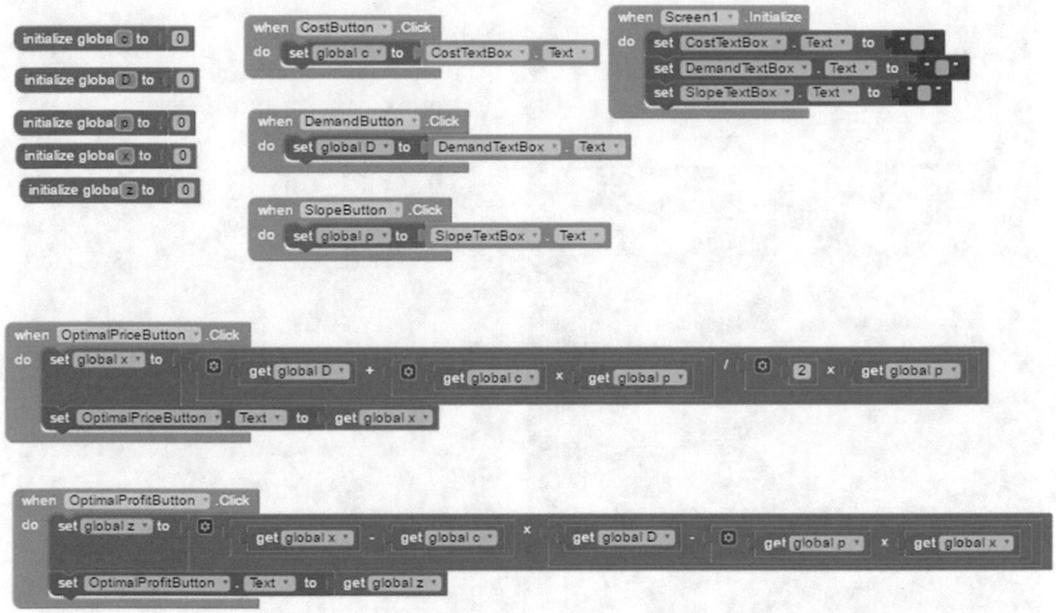

Fig. 5.13 Complete Blocks of the Price-is-right App

5.4.5 Demonstration

Figure 5.14 illustrates the home screen and the final solution for an example problem.

5.4.6 Evaluation

The Price is Right App is very simple, but that is because we have used the calculus to find the key equations. The next App is a concerns an engineering design problem and shows how we can utilize App Inventor in a nonlinear equation environment.

5.5 Speed-Check Ramp Design

William Andrews of the MIE 379 class of 2015 programmed this App to help design ramps for different sports events.

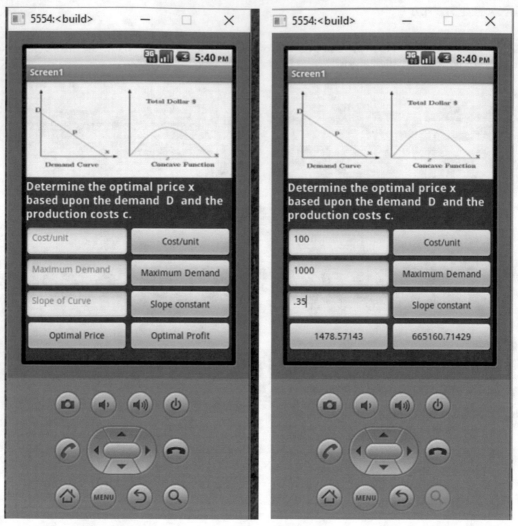

Fig. 5.14 Price is Right App

5.5.1 Introduction

Fig. 5.15 Truck Ramp

This is a good example of solving a problem requiring nonlinear equations which AI2 is capable of doing in a very effective manner. The App is very portable and could be used in a variety of settings where quick calculations can reveal the feasibility of the problem solution.

Across a variety of sports and events (such as rally car racing, skiing & snowboarding, skateboarding, etc.) ramps are used to launch athletes and performers into the air to clear obstacles, distances, and perform aerial maneuvers. Designers of these ramps often face certain constraints, such as a minimum horizontal or vertical distance, the angle of the ramp design, or the speed that can be reached before the ramp launch (Figure 5.15).

5.5.2 Problem

The Speed-Check App allows ramp & course designers to determine how much speed will be needed to clear a given distance or height with their current ramp angle. From this data, they can make appropriate adjustments to the course and ramps. Alternatively they can use Speed-Check to determine the ramp angle necessary to clear a given distance with a given speed constraint.

Figure 5.16 illustrates the designer screen input with all the design variables which we must resolve.

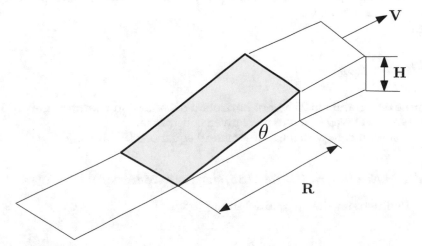

Fig. 5.16 Ramp Design Input

5.5.3 Mathematical Model

While we do not have an objective function per se, the solution of the nonlinear equations is still a challenging problem to generate feasible solutions. Nevertheless, generating a variety of alternative solutions depending on varying the parameters could allow one to finally rank order the alternative designs.

Decision Variables:

R Horizontal Distance in Feet
H Vertical Distance in Feet
θ Ramp Angle in Degrees
V Speed off Ramp in MPH

Assumptions: Wind resistance is neglected.
Constraints:

$$g = 32.15 ft/s \tag{5.36}$$
$$V \geq 0 \tag{5.37}$$
$$\theta \geq 0 \; and \; \theta \leq 90 \tag{5.38}$$
$$R + H \geq 0 \tag{5.39}$$

Nonlinear Equations: *Given Horizontal Distance & Ramp Angle:*
Given Vertical Distance & Ramp Angle:

$$V = ((Rg)/\sin(2*\theta))^1/2 \tag{5.40}$$

$$V = ((2Hg)/(\sin(\theta)^2)))^1/2 \tag{5.41}$$

Given Speed & Horizontal Distance:

$$\theta = (\sin^- 1(((gR)/(V^2))))/2 \tag{5.42}$$

5.5.4 Algorithm

The App successively solves the nonlinear equations to yield the design solution for the ramps. The App's simplicity and portability is a good example of this type of App which can be quite useful in practice.

5.5.5 Solution App

Given sample constraints of 25 feet of horizontal distance and a ramp angle of 35 degrees, a speed of 19.9 MPH was found to meet given constraints.

This checks out nicely with the mathematical equations (Figure 5.17).

$$V = (.681818 mph/(ft/s)) * (((25 ft * 32.15 ft/s)/\sin(2*35 degrees))^1/2) = 19.94 MPH \tag{5.43}$$

Figure 5.18 illustrates the optimal solution.

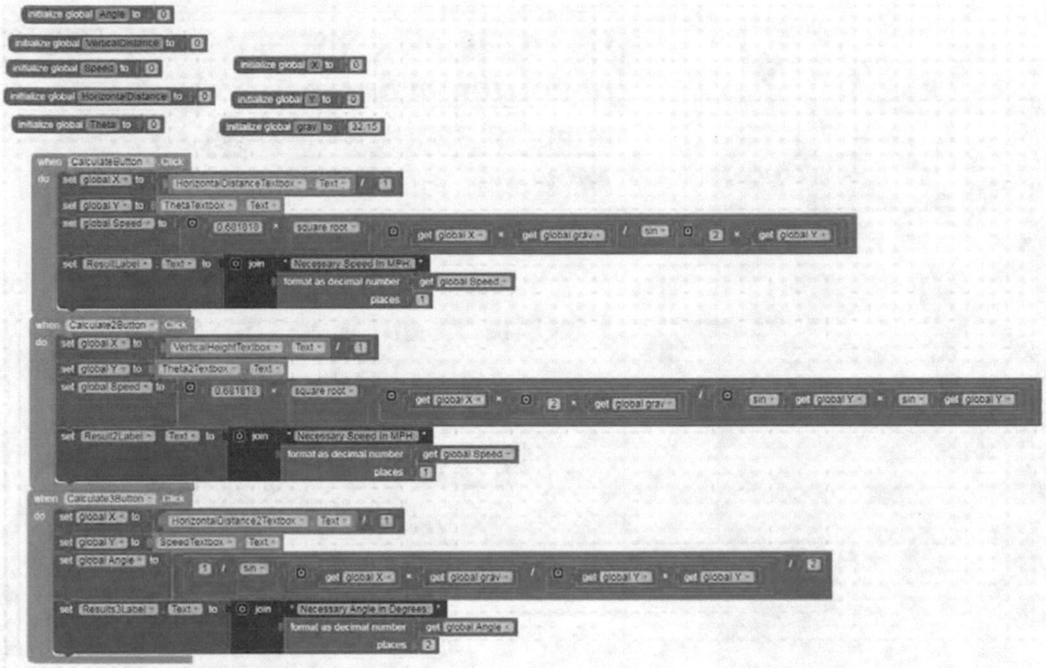

Fig. 5.17 Ramp Design Blocks

5.5.6 Evaluation

While designing ramps and course layouts in the field, the Speed-Check application provides designers quick and simple optimal values for the various constraints involved in designing jumps/ramps. Since projectile motion does not involve mass values (gravity acts on all objects equally) the application can be used for a variety of sports and events, from ski jumps to rally car jumps. The next App concerns a classic problem in Industrial Engineering and Operations Research relying on a good understanding of NLP.

5.6 Weber Location Problem: Weizfeld's Algorithm

This is a classical nonlinear location problem with Euclidean distance and is the counterpart to the rectilinear distance location problem discussed in Chapter 3 called the Pinball Weber problem.

5.6.1 Introduction

We want to demonstrate a dynamic location problem where the inputs are put in by the user and the App computes the optimal location of a new facility after the inputs. Also, of some import, the App demonstrates how multiple lists are searched in AI2.

Let's say we are working out of doors and need to determine the location of a new water well in relation to some existing water wells. The phone App environment is most suitable for this type of situation. We have a map of the area and we want to locate the existing well with Cartesian coordinates and then determine the new well in relation to the existing coordinates. As in the previous factory location problem, we have a given set of demand

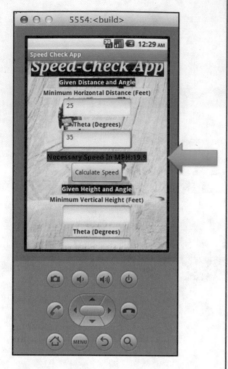

Given sample constraints of <u>25 feet</u> of horizontal distance and <u>a ramp angle of 35 degrees</u>, **_a speed of 19.9 MPH_** was found to meet given constraints.

This checks out with the mathematical equations.

V = (.681818 mph/(ft/s))*(((25 ft * 32.15 ft/s)/sin(2*35 degrees))^1/2)
  =**19.94 MPH**

*This is just one of three calculations that Speed-Check application is capable of.

Fig. 5.18 Ramp Design Example

points and we are interested in locating a new facility in relation to the given facilities such that the unweighted distance from the new facility to the existing facilities is minimized.

5.6.2 Problem

The classical version of the problem is now defined. Given three points in the plane, find a fourth point $s : (X, Y)$ such that the sum of its distances to the other three points is a minimum (Fermat, 17th century):

$$Minimize| \; Z = f(\mathbf{x}) = \sum_{i=1}^{3} w_i d_2(i, j) \tag{5.44}$$

where $w_i =$ flow from points i to points j

$$d_2(i, j) = [(X - a_i)^2 + (Y - b_i)^2]^{1/2} \tag{5.45}$$

P_i has (a_i, b_i) coordinates. The above is an "unconstrained" nonlinear programming problem, often referred to as the Steiner/Weber problem.

We actually have already seen this problem once before in the Pinball Weber Location App in Chapter 3 on Linear Programming, but in this instance, we will develop a nonlinear programming algorithm approach since we are using Euclidean distance directly.

5.6.3 Mathematical Model

There can actually be more than three given facilities and our App will allow for this. We assume Euclidean distance is used to travel between the new facility $i(x,y)$ and the j–existing facilities located at points (a_i, b_i)

$$\text{Minimize Z} = \sum_j w_j dist(i,j) \tag{5.46}$$

We assume differentiability here. So we will also have an unconstrained optimization problem in two variables (X, Y). Classical Calculus:

$$\frac{\partial Z}{\partial X} = \frac{\partial Z}{\partial Y} = 0 \tag{5.47}$$

the above are necessary conditions for the optimality of point s. If $f(\mathbf{x})$ is convex, then the above conditions are also sufficient to guarantee optimality.

$f(\mathbf{x})$, however, is not separable in X, Y.

As mentioned before, it is difficult to directly solve $\frac{\partial Z}{\partial X}$, $\frac{\partial Z}{\partial Y}$ because they are interdependent nonlinear equations and the square roots create round-off error as well as computation problems.

5.6.4 Algorithm

The algorithm we will follow is called Weiszfeld's algorithm. Partial derivatives are given by the following set of equations:

$$\frac{\partial Z}{\partial X} = \sum_{i=1}^{m} \frac{w_i(X - a_i)}{[(X - a_i)^2 + (Y - b_i)^2]^{\frac{1}{2}}} = 0 \tag{5.48}$$

$$\frac{\partial Z}{\partial Y} = \sum_{i=1}^{m} \frac{w_i(Y - b_i)}{[(X - a_i)^2 + (Y - b_i)^2]^{\frac{1}{2}}} = 0 \tag{5.49}$$

If we set the first equation to zero and perform some algebra:

$$X \sum_{i=1}^{m} \frac{w_i}{[(X - a_i)^2 + (Y - b_i)^2]^{\frac{1}{2}}} = \tag{5.50}$$

$$\sum_{i=1}^{m} \frac{w_i a_i}{[(X - a_i)^2 + (Y - b_i)^2]^{\frac{1}{2}}} \tag{5.51}$$

If we further let:

$$g_i(X, Y) = \frac{w_i}{[(X - a_i)^2 + (Y - b_i)^2]^{\frac{1}{2}}} \forall i \tag{5.52}$$

We get the following:

$$X = \frac{\sum_{i=1}^{m} a_i g_i(X, Y)}{\sum_{i=1}^{m} g_i(X, Y)} \tag{5.53}$$

and for the Y-coordinate:

$$Y = \frac{\sum_{i=1}^{m} b_i g_i(X, Y)}{\sum_{i=1}^{m} g_i(X, Y)} \tag{5.54}$$

Thus, if we start from a given point (usually the center of gravity) then we can compute the next point iteratively as (Figure 5.19):

$$X^{k+1} = \frac{\sum_{i=1}^{m} a_i g_i(X^k, Y^k)}{\sum_{i=1}^{m} g_i(X^k, Y^k)} \qquad (5.55)$$

and for the Y-coordinate:

$$Y^{k+1} = \frac{\sum_{i=1}^{m} b_i g_i(X^k, Y^k)}{\sum_{i=1}^{m} g_i(X^k, Y^k)} \qquad (5.56)$$

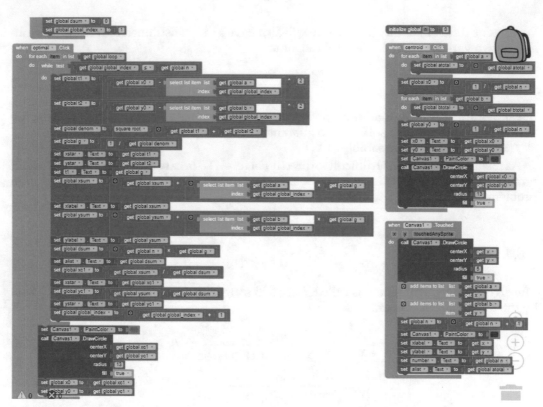

Fig. 5.19 Partial Blocks of the Weiszfeld Location App

5.6.5 Demonstration

The designer screen has the origin in the upper left hand corner. The USGS map is an area in part of eastern Massachusetts. The user inputs the App on the screen and the App records the (x, y) coordinates as they are entered. Figure 5.20 illustrates the input and final optimal solution for an example 3-point problem. The purple circle is the starting solution while the orange circle is the final optimal solution after five iterations. More iterations are possible, but five was a test run for the App. Clicking the Reset button on the App will run another problem input by the user.

Fig. 5.20 Weiszfeld Location App

5.6.6 Evaluation

The program is pretty efficient and works well. Another related problem is the smallest enclosing sphere problem which is treated in the next App section.

5.7 Smallest Enclosing Sphere Problem

A classic geometric location optimization problem is the minimum enclosing circle or sphere problem. Wendy (Xi Jiang) from Smith College who took our class in 2015 programmed the App. She was interested in finding the best party location for four friends in the Pioneer Valley.

5.7.1 Introduction

There are a number of different algorithms for its solution and it maintains an intuitively appealing understanding.

There are many practical applications.

- Find the location of a hospital or post office to service a community population so that the largest distance to the facility is minimized,
- In the military, this is known as the "smallest bomb problem" for obvious reasons.
- Jung developed a theorem which states that every finite set of points with geometric span has an enclosing circle with radius no greater than $\frac{d}{\sqrt{3}}$. So the problem has a solution, in fact, there is an optimal solution.

5.7.2 Problem

This is a classic optimization problem *a.k.a* as the *MinimumEnclosingSphere(Ball)problem*. We are given a set of randomly generated set of points S in the plane/space and we wish to find the ball(circle) B of smallest diameter such that all points in S are or higher dimensions either contained in B or are on its boundary. It has many applications:

Figure 5.21 illustrates three alternative spheres enclosing a common point set. We would like to find the smallest enclosing sphere.

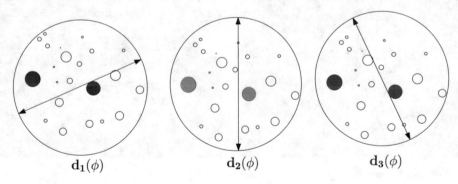

$$\mathbf{d_1}(\phi) \qquad\qquad \mathbf{d_2}(\phi) \qquad\qquad \mathbf{d_3}(\phi)$$

Fig. 5.21 Smallest Enclosing Sphere Alternatives

5.7.3 Mathematical Model

Let the coordinates of the existing points P_i be denoted as $(a_i, b_i) i = 1, \ldots, 4$ and those of the desired point P be given as (x,y). Define $d(x,y) = \max(d(P, P_i)), i = 1, \ldots 4$. Then $d(x,y) \geq d(P, P_i)$, equivalently

$$d(x,y)^2 \geq (x - a_i)^2 + (y - b_i)^2, i = 1,2,3,4 \tag{5.57}$$

By defining a new variable $\lambda = x^2 + y^2 - d^2$, the problem is reduced to the following quadratic programming problem:

$$\text{Minimize } f(\lambda, x, y) = x^2 + y^2 - \lambda \tag{5.58}$$

$$\text{s.t.} \quad 2a_i x + 2b_i y - \lambda \geq a_i^2 + b_i^2, i = 1,2,3,4 \tag{5.59}$$

This quadratic programming problem can then be solved on the NEOS server. Figure 5.22 illustrates the blocks programming.

5.7.4 Algorithm

Because of the nonlinearities involved, she used the NEOS server to solve the problem.

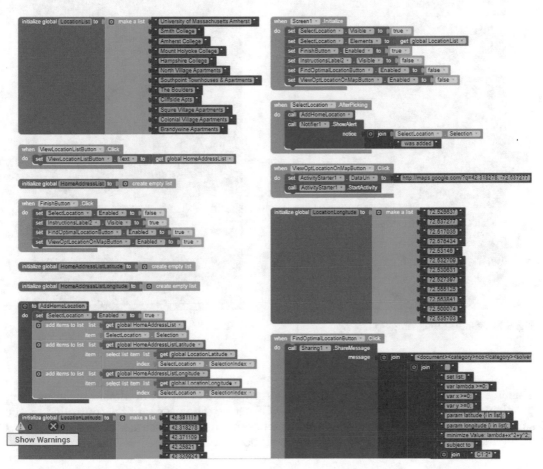

Fig. 5.22 Smallest Enclosing Sphere App

5.7.5 Demonstration

Each of the alternative possible locations is encoded in a list structure within the App with their latitude and longtitude locations. These locations are then sent to the NEOS server along with the optimization problem formulation and eventually solved by the software code MINOS. Figure 5.23 illustrates the simple interface and the alternative locations possible.

The message received from the NEOS server on the phone or tablet yields the optimal location for the meeting point as shown below (Figure 5.24).

The App will also display a Google Map with a picture of the optimal location which is quite useful.

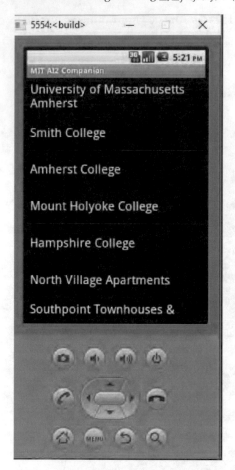

Fig. 5.23 Smallest Enclosing Sphere App Demo

5.7.6 Evaluation

This is a very nicely designed and compact App. It is a complex nonlinear programming problem but it works very well with the NEOS server. The user input is controlled by the sample list of locations, but this is eminently workable and demonstrates that more locations or interactively inputting the coordinates is all possible.

5.8 Cold Storage Warehouse Nonlinear Design Problem

This is an interesting nonlinear optimization problem that involves the use of Lagrangean relaxation to solve it. It is a classical type of constrained nonlinear programming problem.

5.8.1 Introduction

Many nonlinear programming problems involve the determination of optimal dimensions of an engineering design where the area and volume elements lead to a natural nonlinear programming formulation.

```
neos@neos-server.org
11:44 AM (1 hour ago)
 to me
File exists
You are using the solver minos.
Executing AMPL.
processing data.
processing commands.
Executing on prod-exec-5.neos-server.org
3 variables:
 2 nonlinear variables
 1 linear variable
4 constraints, all linear; 12 nonzeros
4 inequality constraints
1 nonlinear objective; 3 nonzeros.

MINOS 5.51: optimal solution found.
3 iterations, objective 1766.752666<-------
Nonlin evals: obj = 6, grad = 5.
x = 21.1591 <-----------------------------
y = 36.3186 <-----------------------------
```

Fig. 5.24 Neos Return Message with Optimal Location Coordinates

5.8.2 Problem

Let's say we are designing a warehouse in a cold climate and we wish to design the warehouse to minimize energy loss. The square foot area heat loss through the sides, walls, and roof say is q times the heat loss through the floor. We have a fixed volume we are trying to satisfy exactly, and we wish to determine the x, y, z coordinates to minimize the overall heat loss of the building. What are the x, y, z dimensions of the building?

5.8.3 Mathematical Model

So we must account for the area of the sides, walls, and roof and relate it to the heat loss through the floor area. Our objective becomes

$$\text{Minimize } Z = q * (2yz, +2xz + xy) + xy \tag{5.60}$$

and the volume constraint is $xyz = v$

5.8.4 Algorithm

When we have a constrained nonlinear programming problem with equality constraints, then a natural algorithm for solution is a Lagrangean relaxation. We will take the equality constraint and put it into the objective function along with a Lagrange multiplier. The Lagrange multiplier is an extra piece of information that is useful for sensitivity analysis, but we will not go into that in any detail. If the problem is not too large, then a symbolic solution may be possible.

The Lagrangean formulation is

$$F := q * (2*x*y + 2*x*z + 2*y*z) + x*y + L*(x*y*z - v) \tag{5.61}$$

and once we differentiate the Lagrangean with respect to x, y, z, we set the derivatives equal to zero and solve, so that we can find the dimensions are equal to

$$x = 2^{1/3} \left(\frac{qv}{q+1} \right)^{1/3} \tag{5.62}$$

$$y = \frac{qv 2^{1/3}}{(q+1)(\frac{qv}{q+1})^{2/3}} \tag{5.63}$$

$$z = \frac{(q+1) 2^{1/3} (\frac{qv}{q+1})^{1/3}}{2q} \tag{5.64}$$

The nice thing about the solution is that we have a symbolic one which is quite general and flexible (Figure 5.25).

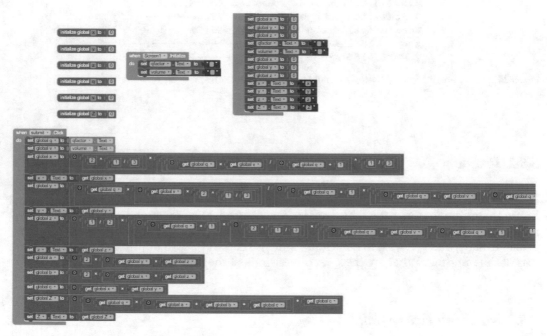

Fig. 5.25 Warehouse Optimization Blocks

5.8.5 Demonstration

Figure 5.26 illustrates the final App design with an example demonstration where we have a given $q-$ factor of 5 and a volume of $75,000$ cubic feet and the solution is $x = 50, y = 50, z = 30, Z = 45,000$. The App works very quickly even with the cube root equations and does not seem to be hampered by large numbers.

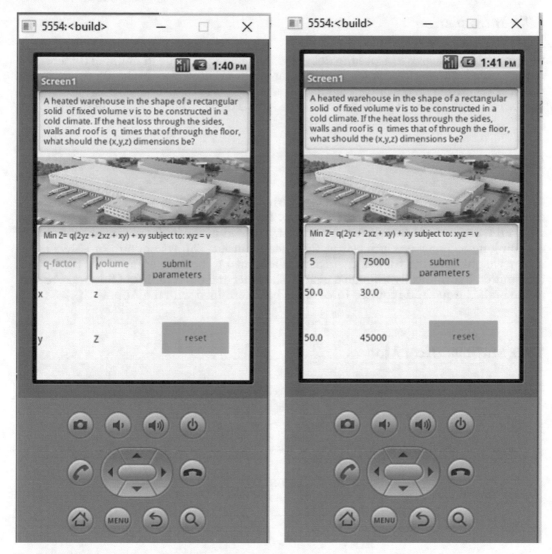

Fig. 5.26 Warehouse Design Optimization Apps and Solution

5.8.6 Evaluation

This is a straightforward App that solves a complex nonlinear programming problem with relative ease. The equations solver works very well even with the complex cube root calculations.

5.9 Disc Brake Engineering Design

This is an example of an engineering design project where a number of pieces of data must be brought together to optimize an engineering system. Normally, these projects require a set of nonlinear equations to be solved so that is why this project is in the NLP category. Luyi Wang a junior in the mechanical engineering department programmed the App.

5.9.1 Introduction

Multiple disk brakes can be used to deliver extremely high torque in minimal dimensional requirements. The multiplication of surface area in a multiple disk brake allows for one of the smallest torque to size ratios available. Several factors such as inner radius, outer radius, and brake torque have to be determined. The App is designed to solve for these quantities.

5.9.2 Problem

A multiple disk brake is driven by a Direct drive brushless motor RBE, and the motor type can be selected from a list picker in the App. The App contains a database of motor types which the user selects from a drop down list. The end of the brake is connected to a rack and pinion mechanism which acts as the system output. The system output is 1.75 of the motor weight, and the friction coefficient is assumed to be 0.20. Meanwhile, the force is transmitted hydraulically through a piston. The inner and outer radius of this brake as well as the brake torque and required force are to be determined with the App.

5.9.3 Mathematical Model

Figure 5.27 illustrates the disk brake system relationships.

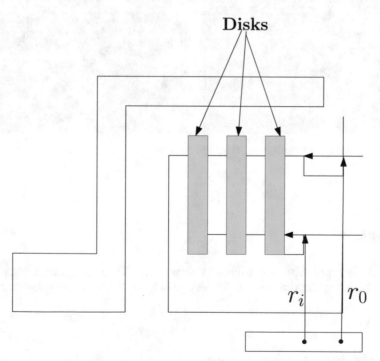

Fig. 5.27 Disk Brake System Diagram

The forces involved in the disk brake system which are important for the phone App are the following: Total normal force acting on the brake:

$$F = \int 2\pi prdr = \pi p(r_0^2 - r_i^2) \tag{5.65}$$

$$\text{Brake Torque} : T = \int_{r_i}^{r_0} 2\pi p_{max} r_i NF_r dr = \pi p_{max} r_i f(r_0^2 - r_i^2)N \tag{5.66}$$

$$\text{Torque, force and outside radius relations} : F = T/f1.58r_0 N \tag{5.67}$$

5.9.4 Algorithm

Figure 5.28 lists a sample of the blocks for the disk brake optimization. There are several lists of input data crucial to the App optimization. Multiple screens are used to help with the input data process.

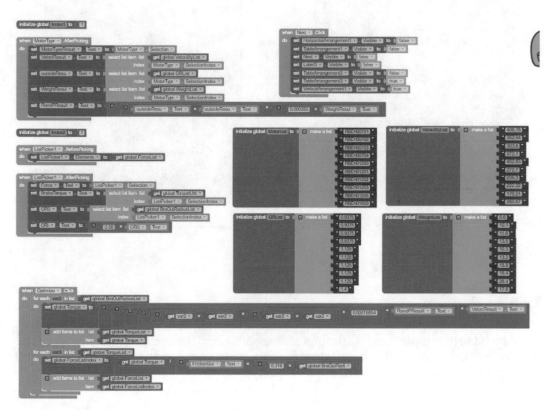

Fig. 5.28 Disk Brake Optimization Blocks

5.9.5 Demonstration

Figure 5.29 illustrates the input process for the brake design and the solution for the Torque and the radius decision variables for the particular motor type. Once the motor type is specified one has to input the Friction Surface and then the Brake Outer Radius, then the App will optimize the remaining values.

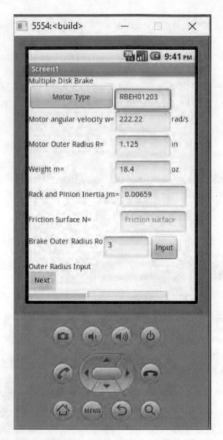 

Fig. 5.29 Disk Brake Example Optimization Apps

5.9.6 Evaluation

The App is fairly complex with all the input information and the database of motor types and the complex calculations that must be carried out. It is a very good example of a successful App for a very complex engineering design problem.

5.10 Farm Crop Planting Problem

This is an example of a farm crop planting problem using Nonlinear Programming (NLP) on the NEOS server to solve the optimal mix of crops. This is a classical operations research problem. It is an ambitious project but eminently suitable for AI2.

5.10.1 Introduction

This is an example of resource planning on a large scale. The App was created by Radha Dutta in 2015 who was a mathematics major in the Mie 379 course.

5.10.2 Problem

This is a very important and challenging problem and of immense practical usefulness. The data were gathered from the UMass Student Farm program.

5.10.3 Mathematical Model

Radha formulated a nonlinear programming problem. Some of the notation are

$\rho_i :=$ Price that crop i sells for.
$\delta_i :=$ Total demand for crop i.
$\gamma_i :=$ crop yield/foot.
$\mu_i :=$ Seeds or plants per foot.
$\sigma_i :=$ Number of seeds per packet.
$\omega_i :=$ Price of seed packet.
$\beta :=$ Total budget available.
$N :=$ Number of crops.

$$\text{Maximize } Z = \sum_i^N \rho_i x_i \tag{5.68}$$

$$s.t. \quad \sum_i^N \varepsilon_i x)i \le \beta_i \text{ where } \varepsilon_i = \left(\frac{2x_i \mu_i}{\gamma_i 640 \sigma_i} \right) \tag{5.69}$$

$$\rho_i, \delta_i, \omega_i, \mu_i, x_i, \varepsilon_i \ge 0 \tag{5.70}$$

5.10.4 Algorithm

Part of the blocks programming is shown in Figure 5.30.

5.10.5 Demonstration

There are five screens associated with the input process and all the parameters. Three of the screens are shown in Figure 5.31.

Data for an example problem are given below

$$
\begin{array}{l|ccccccc}
CropName & Price & Demand & \begin{array}{c} Yield/ \\ foot \end{array} & \begin{array}{c} Seeds/ \\ foot \end{array} & \begin{array}{c} Seeds/ \\ Packet \end{array} & \begin{array}{c} Price/ \\ Packet \end{array} & Epsilon \\
\hline
Beets & 1 & 1690 & 1.25 & 6 & 1850 & 8.4 & 6.81081x10^{-5} \\
Kale & 1.75 & 8600 & 1 & 1 & 3500 & 12.5 & 1.1607x10{-5} \\
Spinach & 8 & 1050 & 0.4 & 12 & 2240 & 9.65 & 0.000403878 \\
Radish & 2 & 875 & 0.50 & 12 & 10600 & 13 & 9.19811x10^{-5} \\
SweetPotato & 1.5 & 3995 & 3.5 & 1 & 1000 & 75 & 6.69643x10^{-5} \\
Carrots & 1 & 10183 & 1.5 & 16 & 100,000 & 101 & 3.36667x10^{-5} \\
BokChoy & 2 & 720 & 0.75 & 2.5 & 5000 & 24.15 & 5.03125x10^{-5} \\
Cauliflower & 1.5 & 830 & 1 & 1 & 500 & 12.45 & 7.78125x10^{-5}
\end{array}
\tag{5.71}
$$

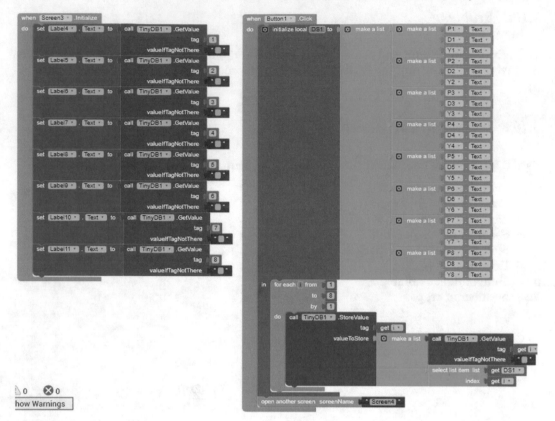

Fig. 5.30 Farm Crop Blocks

The solution achieved from the NEOS server using the MINOS software was for the number of pounds for each crop:

- Beets   = 1690
- Kale   = 13360.50
- Spinach   = 1687.81
- Radish   = 1852.75
- Sweet Potatoes   =3995
- Carrots   =10183
- Bok Choy   =3387.5
- Cauliflower   =1642.57
  Objective Value   Z = 67692.30

It required 119 major iterations for the solution. Notice that some of the crops are equal to their demands, but others exceed their demands.

5.10.6 Evaluation

Everything works very well for a complicated App. There is a long input process if one uses all eight crops. One can utilize fewer crops but one must put in zeros in all the cells where

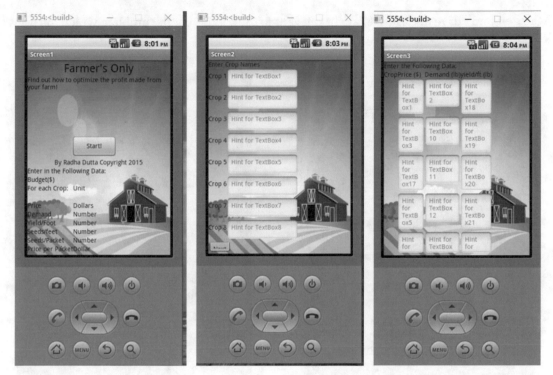

Fig. 5.31 Farm Crop App Screens

numbers are required otherwise the App will not execute on the NEOS server. This is true for most of the Apps in that you must put in dummy answers in all the cells.

5.11 Production Planning and Control

Prashant Meckoni of the Mie 724 class in Nonlinear Programming of 2015 programmed this App. The problem he tackled is pretty complex and his understanding of Nonlinear Programming was essential to its solution.

5.11.1 Introduction

Similarly to the last App, we look at a production planning problem where all the detailed demand and costs of a product are factored into the problem to determine the price/unit and the cost/unit where items have a short shelf life. This is a classical problem in Industrial Engineering and Operations Research with a long history and very important in today's perishable production environment.

5.11.2 Problem

Figures 5.32 illustrates the designer screen input for the App. We seek the optimal price of the perishable product (*e.g.* dairy products) with a short shelf life. There is a lot of data from the demand curve and the cost curve relevant to the solution of the problem which the user must provide.

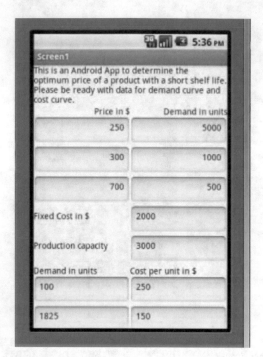 

Fig. 5.32 Input of Production Planning

5.11.3 Mathematical Model

The mathematical model as depicted in Figure 5.33 illustrates the series of constraints in the problem. Figure 5.34 illustrates the blocks.

5.11.4 Algorithm

The objective function is nonlinear for this problem so we must use the MINOS solver of the NEOS site, but the NEOS server will automatically utilize the MINOS module to solve the problem.

5.11.5 Solution App

The solution for the given data yields

$$Z = \$367,128 \text{total cost; price/unit} = \$240.712; \text{cost/unit} = \$117.67 \qquad (5.72)$$

Fig. 5.33 Mathematical Model of Production Planning

5.11.6 Evaluation

Because of the complexity of this problem, we must rely on the NEOS server which actually solves the problem fairly quickly even given the large number of constraints.

5.12 Erlang Loss M/G/c/c Ambulance Staffing App

This is a special type of queueing problem that has a great number of applications in telecommunications, industrial engineering, and operations research, and other related engineering and service system topics.

5.12.1 Introduction

This queueing problem is one of the first queueing problems ever studied. It was originally started by A.O. Erlang for the Copenhagen Telephone Exchange in 1909 who was interested in determining the number of human operators to manage the telephone switch at the exchange. The formula he arrived at is the basis for the mathematical model of this App.

5.12.2 Problem

For many systems, an arrival who finds all servers occupied is for all practical purposes lost to the system. We would like to minimize the probability of lost customers.

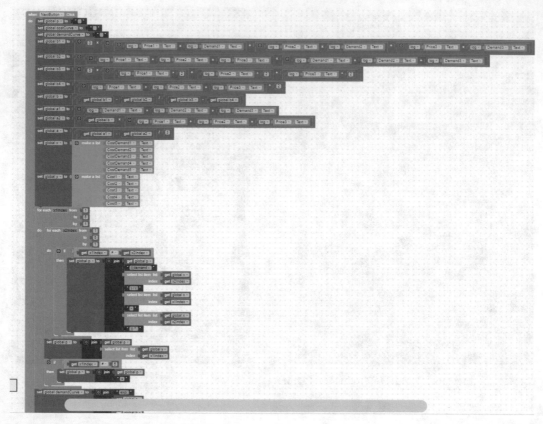

Fig. 5.34 Production Planning Blocks

- Suppose we call an airline reservations system and all reservations clerks are busy.
- Suppose some one calls for an ambulance to respond to an accident or other emergency, and there are no ambulances available to respond?

Obviously, the ambulance problem is most serious. How can we insure that we have enough ambulances for an urban area to minimize this unfortunate circumstance.

5.12.3 *Mathematical Model*

We have random arrivals (*i.e.* Poisson arrivals) and random service from a general probability distribution and a finite number of servers c and finite waiting room actually equal to the number of servers (c). This is called an $M/G/c/c$ queue for short.

The Erlang formula is

$$P_c = \frac{(c\rho)^c/c!}{\sum_{i=0}^{c}(c\rho)^i/i!}, \rho = \lambda/c\mu \tag{5.73}$$

5.12.4 Algorithm

A recursive procedure based upon a nonlinear equation is utilized to solve the problem where we take the input information and systematically find the threshold probability which satisfies the requirement for the system problem.

The blocks programming is shown in Figure 5.35.

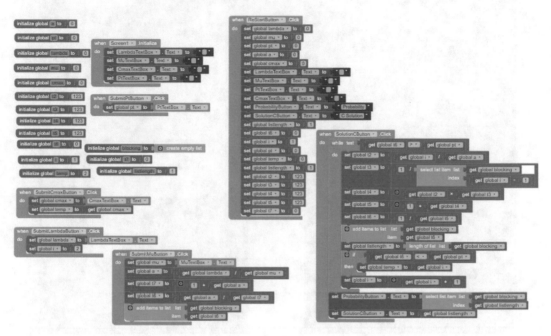

Fig. 5.35 M/G/c/c App Programming

5.12.5 Demonstration

Let's say that we are trying to plan the number of ambulances for an urban area. The idea for this example application came from a problem suggested by Winston [15]. Suppose that on average 21 calls per hour are received at the 911 service center. An ambulance responding to a call takes on average 25 minutes to pick up a patient and deliver the patient to the hospital. The ambulance is then able to respond to other emergency calls. How many ambulances should the hospital maintain in order to ensure that there is at most a 0.01 probability of not being able to respond immediately to an emergency call? The App used for the input and the final solution are shown in Figure 5.36. The App was also used to determine the number of tables needed in a restaurant.

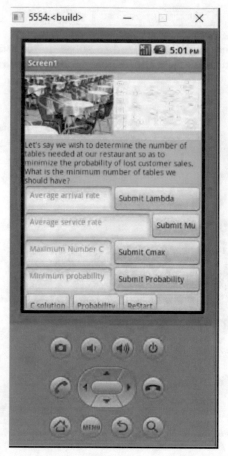 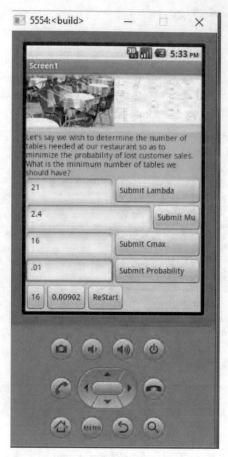

Fig. 5.36 M/G/c/c App Screens for Ambulance Problem

5.12.6 Evaluation

The program works very well and is very easy to use.

5.13 Economic Order Quantity (EOQ) Problem

Inventory problems are classical optimization problems and have a glorious well-known
history in OR. They are also the basic problem of most supply-chains.

Most businesses must maintain a certain amount of inventory of a product in order to
meet demand for a product which is often uncertain. If there is too much inventory, then
the holding costs can become significant, otherwise, if there is not enough inventory, then
shortage costs are incurred. Inventory is replenished periodically at a certain setup cost and
a unit cost is incurred for the product.

5.13.1 Introduction

Let's say that we have a manufacturing company that must supply R parts at a constant
rate during planning period T. Demand is fixed and known and we will assume that no

shortages are permitted. Figure 5.37 illustrates the situation graphically [3] of the saw-tooth inventory problem.

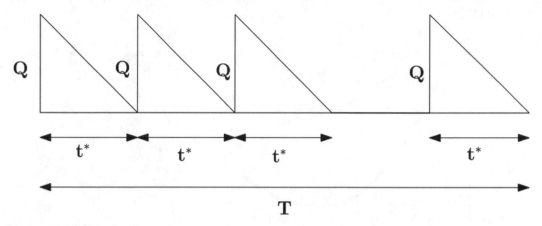

Fig. 5.37 EOQ Objective Function

If we itemize the costs involved in an inventory system, we have the following general equation [14]:

$$\text{Total Cost} = \text{Purchasing\$} + \text{Setup\$} + \text{Holding\$} + \text{Shortage\$} \tag{5.74}$$

Thus, the basic questions are

- **How much** should we order?
- and **When** should we order?

In symbols, we have

Q:= Economic order quantity in pieces.
C:= Unit costs in dollars per piece.
C_1:= Holding costs per piece in dollars.
C_s:= Reorder or set-up costs in dollars.
R:= Demand for the product.
T:= Total Time per planning period in daya.
t^* := Time in days between production runs.

Let Q represent the run size and T^* the interval of time between runs, then R the total demand for the planning period is

$$\frac{R}{Q} := \text{number of runs during time } T \tag{5.75}$$

Hence,

$$t^* = \frac{T}{R/Q} = \frac{Tq}{R} \tag{5.76}$$

If the interval t^* begins with Q units in stock and ends with non, then:

$$\frac{Q}{2} := \text{average inventory level during } t^* \tag{5.77}$$

$$\frac{Q}{2}C_1 t^* := \text{inventory costs during } t^* \tag{5.78}$$

If we graph the functions involved in the problem we are led to the following characterization of the problem where the lower curve represents the setup costs and the linear relation the carrying costs, and the composite curve on top is the total expected costs of the problem [17] (Figure 5.38).

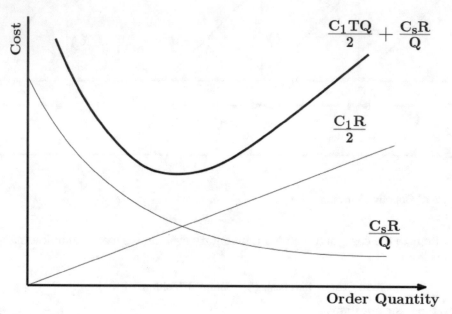

Fig. 5.38 EOQ Objective Function

5.13.2 Problem

We would like to find the optimal order quantity Q and the time between production runs t^*.

5.13.3 Mathematical Model

For the EOQ model, we assume there are no shortages allowed. There are many variations of the basic model, see [3]. We will follow the development of Churchman and Ackoff [3] and that of Woolsey and Swanson [17] since it incorporates the planning horizon and also allows us to calculate the optimal lead time. Combining all our symbols and the basic equations together, we get the following objective function:

$$TEC = \frac{C_1 TQ}{2} + \frac{C_s R}{Q} \tag{5.79}$$

Differentiating the function with respect to Q and setting the equation to 0, we find that

$$Q^* = \sqrt{\frac{2C_s R}{C_1 T}} \tag{5.80}$$

Furthermore,

$$t^* = \sqrt{\frac{2C_s T}{C_1 R}} \tag{5.81}$$

and the Total Expected Cost (TEC) is

$$TEC^* = \sqrt{2C_1 T C_s R} \tag{5.82}$$

5.13.4 Algorithm

The algorithm is based upon the use of the calculus for determining the equations for the optimal quantities of Q, T^*. The algorithm is a fairly straightforward implementation of AI2 blocks after reading in the parameters of the equations. Figure 5.39 illustrates all the blocks for the EOQ model. Aside from the input quantities, one can see the three sets of blocks needed to compute the optimal quantities, Q^*, t^*, TEC^*.

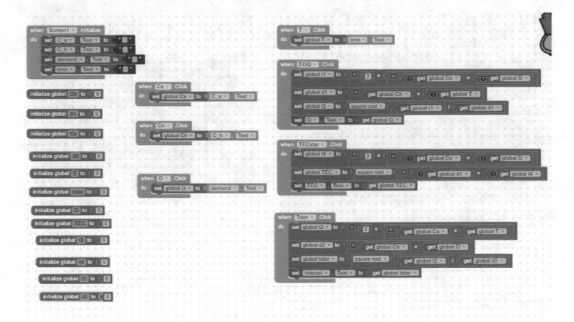

Fig. 5.39 EOQ Blocks

5.13.5 Demonstration

Figures 5.40 reveals the input data screens and the outputs values of the parameters of the example problem.

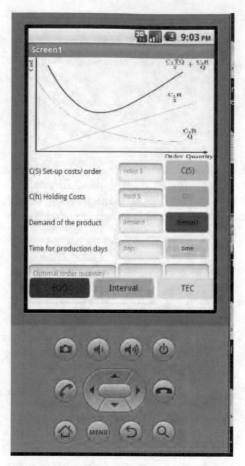 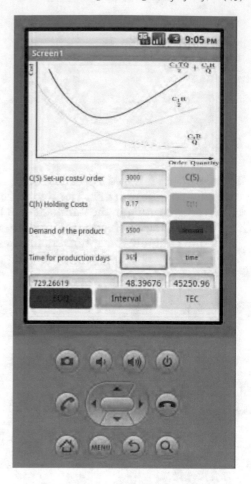

Fig. 5.40 EOQ Example Problem

5.13.6 Evaluation

The App works very well, is fast and efficient. Extensions to this model are eminently possible, see [3], and are most valuable.

Finally, we present a Generalized Nonlinear Programming App

5.14 Generalized Nonlinear Programming Problem

In a similar vein to the Linear and Integer programming problems, we present a Nonlinear App for solving general NLP problems.

5.14.1 Introduction

The nonlinearities can be in the objective function or constraints. The success in solving these problems will depend upon the convexity and non-convexity of these problems.

5.14.2 Problem

The general dilemma for NLP problems is that there is no simplex-type algorithm for their solution. The local vs. the global nature of NLP rears its head in these situations.

5.14.3 Mathematical Model

The general model is as follows:

$$\text{Max or Min } f(\mathbf{x}) \tag{5.83}$$
$$s.t.: \text{ inequalities } g_i(\mathbf{x}) \leq 0 \tag{5.84}$$
$$\text{equalities } h_i(\mathbf{x}) = 0 \tag{5.85}$$
$$\text{nonnegativity } \mathbf{x} \geq 0 \tag{5.86}$$

5.14.4 Algorithm

The NEOS server is the desired medium for the solution, although as we have seen specialized algorithms to capture the special structure of the NLP may be used to our advantage as in the location and network design problems. Figure 5.41 illustrates the programming blocks.

5.14.5 Solution App

Let's first solve the following LP as an LP with a linear objective function. Then let's solve it as an NLP with a nonlinear objective function.

$$Minimize\ Z: -\frac{1}{2}x_1 - 2x_2 \tag{5.87}$$
$$s.t.: \ x_1 + x_2 \leq 6 \tag{5.88}$$
$$x_1 - x_2 \leq 1 \tag{5.89}$$
$$2x_1 + x_2 \geq 6 \tag{5.90}$$
$$-\frac{1}{2} + x_2 \leq 4 \tag{5.91}$$
$$-x_1 \leq 1 \tag{5.92}$$

For the LP, we get the following solution which is at a corner point of the convex hull which makes sense for the LP solution. Since the objective function and the constraint se are convex, this is the global optimal solution.

$$z =, x_1 = \frac{4}{3}, x_2 = \frac{14}{3} \tag{5.93}$$

It is interesting to see the linear objective function contours that pass over the solution set.

For the nonlinear program with the following objective function,

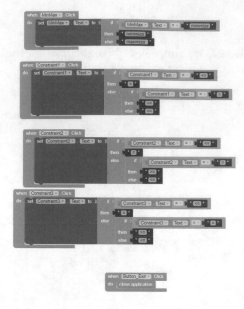

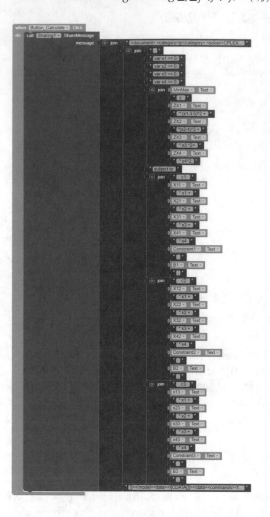

Fig. 5.41 NLP Blocks

$$\text{Minimize } Z : 10(x_1 - 3.5)^2 + 20(x_2 - 4)^2 \tag{5.94}$$

we get

$$z = 15, x_1 = 2.5, x_2 = 3.5 \tag{5.95}$$

The objective function here is a circle centered at $(3.5, 4)$ and the nearest point on the convex hull is the optimal point, which is along an edge of the convex polytope. The solution may occur at an extreme point, along an edge, or anywhere in between. So this is not really expected, but given the circular functions, this turns out to be the global minimum. Again the circular contours are indicative of the nonlinear effects of the objective function. So, in general, with NLP we cannot know for sure where the solution is going to emerge on the solution set, even if it is convex. We also may have to pre-solve the problem with a starting solution (Figure 5.42).

5.14.6 Evaluation

The general NLP App is not the most general App since one must tailor the objective function to the nonlinear relationship between the variables. We have simplified our problem because it is separable in the decision variables and this may or may not always be the case. One

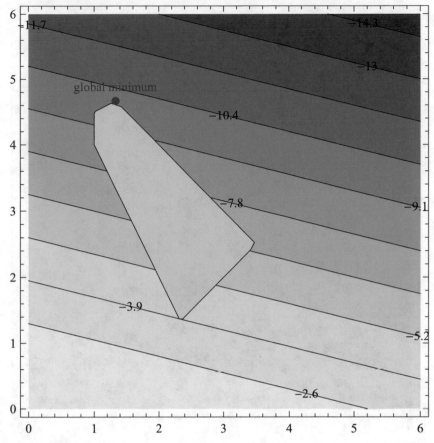

Fig. 5.42 Linear Example Problems

can also try to linearize the problem using (*e.g. logarithms* for the objective function and constraints or use piecewise-linear approximations, but this may not always be possible or desirable. So, NLP problems remain one of the most difficult problems alongside integer programs (Figure 5.43).

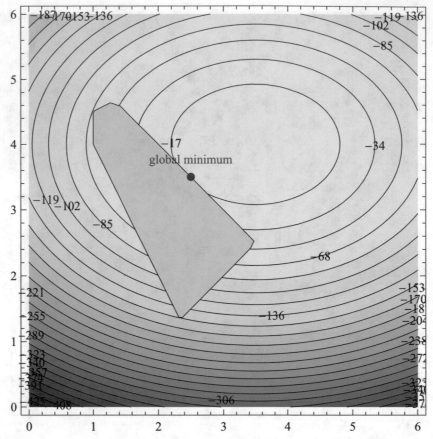

Fig. 5.43 Nonlinear Example Problems

5.15 Exercises

1. **Geometric Problem**

   A rectangular mirror with area A square feet is to have trim along the edges. If the trim along the horizontal edges costs p cents per foot and the trim for the vertical edges costs q cents per foot, find the dimensions of the mirror which will minimize total costs.

2. **Zononi's Garage Problem**

   Suppose it is possible to choose a service rate $\mu > \lambda$ the arrival rate, for Zononi's Garage in order to minimize the cost of repairing vehicles. Let's choose an objective function where

   $$C(\mu) = c_1 * L + c_2(\mu) \tag{5.96}$$

   where c_1 is the waiting cost per customer per unit time and c_2 is the operating cost per unit time as a function of the service rate.

   If we assume for simplicity that $c_2(\mu) = c_2\mu$, a linear approximation, then because from queueing theory, we know that $L = \frac{\lambda}{\mu - \lambda}$ where (L:= average number of customers waiting and in service) we obtain the total cost function:

$$C(\mu) = \frac{c_I \lambda}{(\mu - \lambda)} + c_2 \mu \qquad (5.97)$$

o Find an expression for the optimal service rate μ^*. o Program an App with App Inventor so Mr. Zononi can determine how he can set the service rate μ.

3. **Wireless Location Problem (after Chong and Zak)**

   Suppose we have a simplified cellular wireless system shown below with two base stations. Both antennas transmit signals to the mobile user **x** at equal power. The power of the received signal as measured by the mobile user is the reciprocal of the squared distance from the different antenna (primary or neighboring base station).

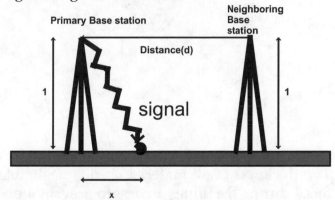

   We are trying to determine the optimal location of a mobile phone **x** to maximize the signal-to-interference ratio, which is the ratio of the received signal power from the primary base station to the received power from the neighboring base station. The squared distance from the mobile to the primary base station is $1 + x^2$, while the squared distance to the neighboring antenna is $1 + (2 - x)^2$. Therefore, the signal-to-interference ratio is

$$f(x) = \frac{1 + x^2}{1 + (2 - x)^2} f'(x) = \qquad \frac{4(x^2 - 2x - 1)}{1 + (2 - x)^2} \qquad (5.98)$$

   Setting the derivative equal to zero, we find

$$x^* = 1 - \sqrt{2} \qquad \text{or } x^* = 1 + \sqrt{2} \qquad (5.99)$$

   It actually turns out that

$$x^* = 1 - \sqrt{2} \qquad (5.100)$$

   is optimal. Develop an App to solve this location problem.

4. **Container Design Problem: (After Pierre)**

   As vice president of a small but dynamic chemical company, you are to investigate containerization for product shipping. Not only should the containers simplify shipment, but also because the containers will be left with the customers, sales should be enhanced. Upon selecting a particular customer as a representative example of the problem, you find that each month 1000 cu.ft. of a chemical must be shipped to this customer.

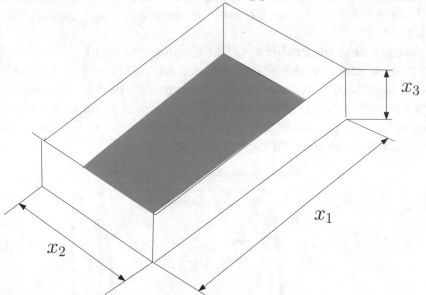

A chemical is to be sent in rectangular containers of length x_1, width x_2, and height x_3. The thickness of all material used is negligible. The container seams are fused during the filling process to prevent moisture from entering. The sides and bottom must be made of scrap metal for which there is no charge. However, only 10 square feet of this scrap can be guaranteed for process per container each month from the materials department, but can be processed to any dimension. Material for the ends cost \$2.00 per square foot, while material for the top cost \$3.00 per square foot. There is also a shipping charge of \$2.00 per container sent.

You must determine how many and what size of containers are needed to ship the chemical at lowest possible cost. Thus the total cost (shipping cost plus the cost of the two ends and the top for

$$\frac{1000}{x_1, x_2, x_3} \tag{5.101}$$

containers) must be minimized subject to scrap limitation. Please formulate as an NLP and solve with AMPL. Once you program the App, you can generalize the parameters to solve for different container designs.

5. **Bridge Design Preliminaries** (After Russel)

Let the set S be described by

$$S = \{x \in E^n : f(\mathbf{x}) \le c\} \tag{5.102}$$

where $f : E^n \to E^1$ is convex and differentiable and $\nabla f(\mathbf{x_0}) \neq 0$ whenever $f(\mathbf{x}) = c$. Let $\mathbf{x_0}$ be a point such that $f(\mathbf{x_0}) = c$ and let $\mathbf{d} = \nabla(\mathbf{x_0})^t$. Let λ be any positive number and define $\mathbf{x_1} = \mathbf{x_0} + \lambda \mathbf{d}$. From the Theorem of the Separating Hyperplane (Bazarra, Jarvis, and Sherali, textbook p. 766), one can show that

$$||\mathbf{x_1} - \mathbf{x_0}||_e < ||\mathbf{x_1} - \mathbf{x}||_e \qquad (5.103)$$

whenever $\mathbf{x}$ is a point of S different from $\mathbf{x_0}$.

Bridge Design Example
Now suppose we have two islands with shapes described by

$$f(x,y) \le c_1 \quad \text{and} \quad g(x,y) \le c_2 \qquad (5.104)$$

where f and g are continuously differentiable and strictly convex functions whose gradient vectors are nonzero when $f(x,y) = c_1$ and $g(x,y) = c_2$, respectively.

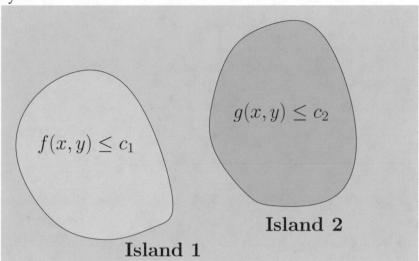

It is desired to build a bridge between the two islands of shortest possible length. Using the result of the *Theorem of the Separating Hyperplane*, devise a search technique to find the optimal bridge location. Test your routine (by hand or computer) on

$$f(x,y) = x^2 + 2y^2, c_1 = 1 \qquad (5.105)$$
$$g(x,y) = (x-4)^2 + (y-3)^2, c_2 = 1 \qquad (5.106)$$

6. **Fire Station location**
   There are three towns assumed located in the Euclidean plane, and for an appropriate scaling and the same origin, their locations are

   $$(-4,8), (6,0), (-2,-5).$$

   A new fire station is to be located to serve the three towns. The towns have respective populations of $4000, 3000$, and 6000. It is desired to locate the new fire station so as to minimize

   $$\sum_i \frac{\text{population of town } i}{\text{total population}} \times (\text{distance to station})^2 \qquad (5.107)$$

Towns 1 and 2 are ruthlessly competing for a new thumbtack factory, each wanting the fire station nearer to it than its rival; so to compromise, the new station must be precisely equidistant from towns 1 and 2. Find the optimal location for the fire station.

AMPL NEOS Link

All's well that ends well

—SHAKESPEARE

A.1 AMPL Programming

A Mathematical Programming Language (AMPL) is a sophisticated software interface platform for linking an optimization problem to a software solver for the actual solution to the problem. It is the linking email software for an AI2 problem to the NEOS server where the solver software resides.

As we know, Ampl can solve much more complex problems than App Inventor can, but App Inventor has the advantage of being portable on a phone or tablet. We can allow App Inventor to solve any problem that we could solve in Ampl if we can send code by email to a server that can do the calculations for us. There is such a server at the Wisconsin Institutes for Discovery at the University of Wisconsin in Madison: the NEOS Server (http://www.neosserver.org/neos/). The NEOS server works by receiving emails of Ampl code formatted in a particular way and then sending a reply email with the answer. App Inventor can send emails and also can be used to allow users to input data into the model. The goal of this instruction set is to explain and provide examples of App Inventor code that will send the correct emails to the server. Technically, we are not going to use emails per se, but xml files sent to NEOS through a Python interface.

The Apps in this section aside from the general Linear Programming (LP) ones, will focus on a Minimum Cost Flow conceptual organization as depicted in Figure A.1.

A.2 General Model

The general model is a simple example that can be used to solve a basic problem with an objective function and constraints. To get started in App Inventor you need to have already written an AMPL code that solves your problem.

Let's say this is the text for your Ampl model:

© Springer Nature Switzerland AG 2021
J. MacGregor Smith, *Combinatorial, Linear, Integer and Nonlinear Optimization Apps*,
Springer Optimization and Its Applications 175,
https://doi.org/10.1007/978-3-030-75801-1_A

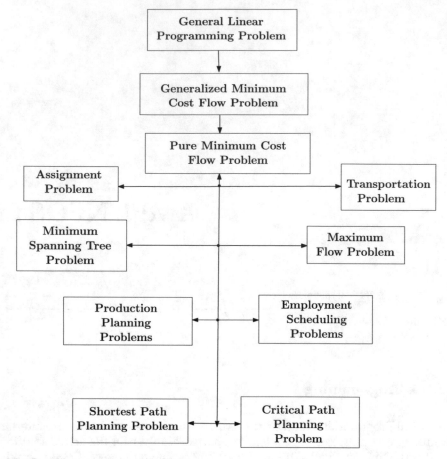

Fig. A.1 Min Cost Flow and Related Problems

$$var\ x1 >= 0; \tag{A.1}$$

$$var\ x2 >= 0; \tag{A.2}$$

$$var\ x3 >= 0; \tag{A.3}$$

$$var\ x4 >= 0; \tag{A.4}$$

$$maximize\ z := 2*x1 + 3*x2 + 4*x3 + 5*x4; \tag{A.5}$$

$$s.t.c1 : 12*x1 + 0*x2 + 13*x3 + 9*x4; <= 97; \tag{A.6}$$

$$c2 : 6*x1 + 4*x2 + 0*x3 + 7*x4; <= 51; \tag{A.7}$$

solve; display z,x1,x2,x3,x4;

You will, in general, take all the information from this model and put it into the basic syntax for the NEOS server to solve it. There are additional pieces of text that the server needs to be able to read the email correctly. The format will be similar to HTML language where there are opening and closing tags on either side of the inputs. There are several ways to interact with the NEOS server. Perhaps the easiest is the Web Interface which we will demonstrate for the MST solution. For the Web Interface, one must create three separate files from our AMPL approach:

- The Model file;
- The Data file;
- The Command file.

Once these separate files are created, one simply inserts them on the Web Interface dialogue, see Figure A.2 and executes the program.

A.3 Minimum Spanning Tree

Lets examine a Minimum Spanning Tree problem. Chapter 2 illustrated a great detail about the MST problem. Here, we let NEOS solve the problem.

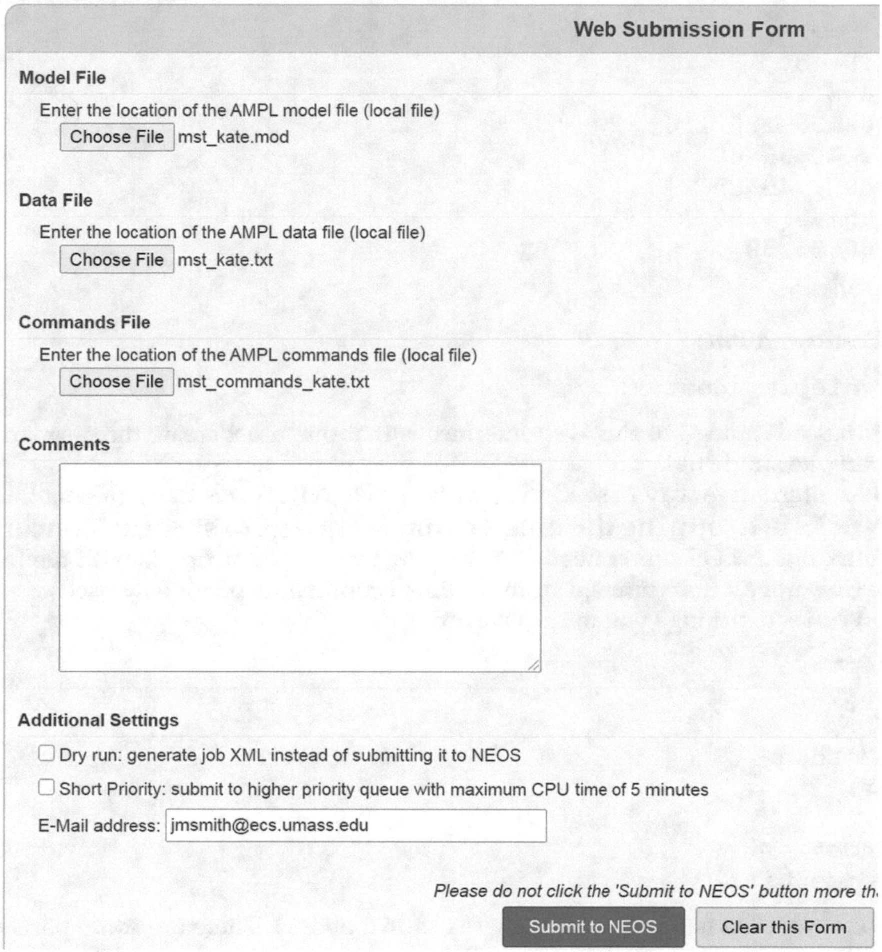

Fig. A.2 NEOS Web Interface

The Model File:

```
set D ordered;
set P within {D,D};
param d {P};
set Adj {i in D} within P := {(u,v) in P: u==i or v==i};
param n := card(D);
set S := 0 .. (2**n-1);
set POW {k in S}:={i in D: (k div 2**(ord(i,D)-1)) mod 2 ==1};
var x {P} binary;
subject to span {i in D}: sum{(u,v) in Adj[i]} x[u,v] >= 1;
subject to size: sum{(i,j) in P} x[i,j] = n -1;
subject to no_cycles{k in S: card({i in POW[k],j in POW[k]:(i,j)in P})>=3}:
sum{i in POW[k], j in POW[k]: (i,j) in P} x[i,j] <= card(POW[k]) - 1;
minimize cost: sum{(i,j) in P} d[i,j] * x[i,j];
```

The Data File:

Notice that the diagonal elements of the distance matrix indicated by a period . are not computed by AMPL.

```
set D := 1 2 3 4 5
;

param :P: d:
1 2 3 4 5 :=
1 . 100 125 120 110
2 100 . 40 65 60
3 125 40 . 45 55
4 120 65 45 . 50
5 110 60 55 50 .
;
```

The Command File:

```
solve; display cost,x;
```

A basic disadvantage of this Web Interface is that one must create three separate files, rather than one single file.

Another alternative way to send the AMPL files to NEOS was through email. Unfortunately, **NEOS is dropping the email connection due to security concerns.** The basic syntax that NEOS server needs for the new way of interaction is with the following template (it is not all that different from the email commands previously used).

As you can see, the tags you must have are

```
<document>,
<category>,
<solver>,
<inputMethod>,
<model>,
<data>,
<commands>,
and <comments>.
```

Each of these also needs a closing tag (as shown above). There are some parts that are already filled in and don't need to be changed for our purposes: category: lp; solver: CPLEX; input method: AMPL (In other applications these could be changed). Here you do not need data and comment. Any numbers or letters that we will input to specify our App will always be placed in a similar way: inside the [] after the relevant "CDATA". We will mostly be concerned with what goes inside the model brackets and the commands brackets. We also break this up into two sections as shown below

```
The first section of code concerns the linking of AMPL with NEOS:
<document> <category>lp</category> <solver>CPLEX</
solver><inputMethod>AMPL</inputMethod> <model><![CDATA[
```

You must do the following on the Display page of your model:

- Add a *notifier button* from the *User Palette* to the Display page.
- Add a **TextBox 1** to this main display page and in the dialogue on the text box type in your email where you want the results response open to be sent. This is handy because before, it was sent to Google Mail automatically.
- Most importantly, go to the *Connections Pallete*, and click on the little sphere **Web icon**. If you click on the icon and drag it into the commands page, a Properties response opens on the right side of the Display and you should put in the following Properties (Figs. A.3 and A.4):

```
Create temp directory /var/lib/condor/execute/dir_13091/neos-13098
File exists
You are using the solver cplexamp.
Checking ampl.mod for cplex_options...
Checking ampl.com for cplex_options...
Executing AMPL.
processing data.
processing commands.
Executing on prod-exec-4.neos-server.org

20 variables, all binary
22 constraints, all linear; 200 nonzeros
        1 equality constraint
        21 inequality constraints
1 linear objective; 20 nonzeros.

CPLEX 12.10.0.0: threads=4
CPLEX 12.10.0.0: optimal integer solution; objective 235
3 MIP simplex iterations
0 branch-and-bound nodes
cost = 235

X :=
1 2   1
1 3   0
1 4   0
1 5   0
2 1   0
2 3   1
2 4   0
2 5   0
3 1   0
3 2   0
3 4   1
3 5   0
4 1   0
4 2   0
4 3   0
4 5   1
5 1   0
5 2   0
5 3   0
5 4   0
;
```

Fig. A.3 MST NEOS Web Output

- 1. In the time slot type 30000.
- 2. In the connection slot type the web command
- 3. Here is what the first few lines should look like

```
The second section of code**:
]]></model><data><![CDATA[]]>
</data><commands>
<![CDATA[solve;display x1, x2, x3,x4,z;]]>
</commands>
<comments><![CDATA[]]>
</comments>
</document>
```

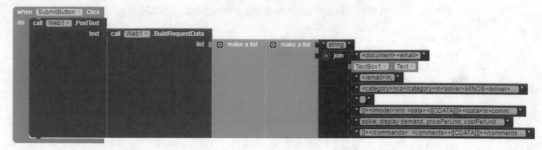

Fig. A.4 NEOS Submission

```
######################################
<document>
<category>lp</category>
<solver>CPLEX</solver>
<inputType>AMPL</inputType>
<client>Mozilla/5.0 (Windows NT 6.1; Win64; x64) AppleWebKit/537.36
  (KHTML, like Gecko) Chrome/83.0.4103.97 Safari/537.36@24.34.20.224</client>
<priority>long</priority>
<email>jmsmith@ecs.umass.edu</email>
<model><![CDATA[]]></model>

<data><![CDATA[]]></data>

<commands><![CDATA[]]></commands>

<comments><![CDATA[]]></comments>

</document>

#########################################
Notice that the first section of code ends right
where you will input the model
(<model><![CDATA[ ).

This makes sense because we need to use additional blocks
 to specify the model using App Inventor.

Also notice that the
 <commands><![CDATA[]]> contains the Ampl model commands.

You will simply copy the exact wording of the
commands from Ampl with no additional spaces
and with the ``;'' after each command.

 The commands that are used in our case are
 ``solve'' and ``display''.
After the display command, make sure to add
any other values that you want displayed at the end
(beyond x1, x2, x3, x4, and z, if any).
```

We shall demonstrate the procedure with the various problem setups (Fig. A.5).

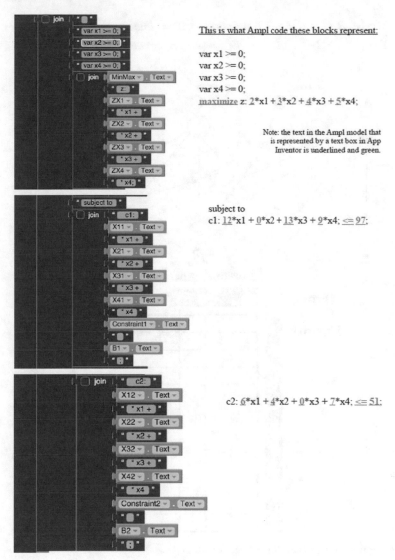

This is what Ampl code these blocks represent:

var x1 >= 0;
var x2 >= 0;
var x3 >= 0;
var x4 >= 0;
maximize z: 2*x1 + 3*x2 + 4*x3 + 5*x4;

Note: the text in the Ampl model that is represented by a text box in App Inventor is underlined and green.

subject to
c1: 12*x1 + 0*x2 + 13*x3 + 9*x4; <= 97;

c2: 6*x1 + 4*x2 + 0*x3 + 7*x4; <= 51;

Fig. A.5 AMPL syntax

A.4 Assignment Problem

The first problem is concerned with the Assignment Problem which has been illustrated many times in the text, see Chapter 2. We will not repeat all the details of the model but focus on the AMPL and NEOS requirements.

A.4.0.1 Introduction

Many decision problems involve the *assignment/matching* of elements of one **set** to elements of another set subject to certain rules. Graphically, we can picture the situation as follows in Figure A.6:

$$X = \{A, B, C, \ldots, Z\} \rightarrow A := \textit{persons}, B := \textit{time slots}, C := \textit{activities} \tag{A.8}$$

$$Y = \{E_1, E_2, E_3, \ldots, E_m\} \rightarrow E_1 := \textit{jobs}, E_2 := \textit{activities}, E_3 := \textit{locations} \tag{A.9}$$

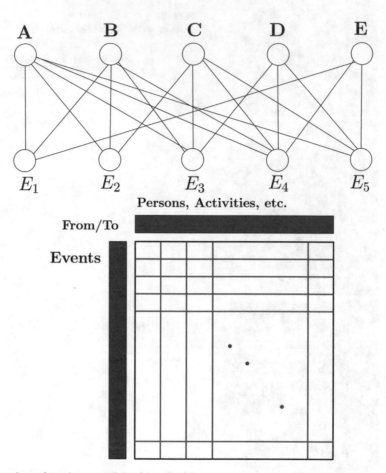

Fig. A.6 Illustration of Assignment/Matching Problem

Normally, there is also an objective function (performance measure or criterion) that evaluates the value of one particular assignment relative to another, and the problem is to choose which of the possible assignments optimizes the value of the objective function.

o   Cost of the assignment ($)
o   Preference of each assignment $(1 \ldots 10; (-5 \ldots 0 \ldots +5))$
o   Maximum reliability of each assignment $(p \in (0, 1))$

A.4.0.2 Mathematical Model

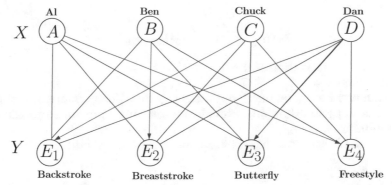

In the language of graph theory, we have a bipartite graph with vertex set

$$V = X \bigcup Y \qquad \text{and edge set} \qquad E \tag{A.10}$$

Each edge e connects a vertex of X to a vertex of Y. Moreover $|X| = |Y|$. For each edge $e \in E$ we are given a nonnegative weight w_e. We want to find a subset $M \subseteq E$ of edges such that each vertex of X and Y is incident to exactly one edge of M and the sum $\sum_{e \in M}$ is a minimum.

A.4.0.3 Problem

Frobenius (1912) and Dénis König (1915) were the first to treat the bipartite matching assignment problem which is a special case of the general Assignment problem.

A.4.0.4 Mathematical Model

In order to form this as an optimization problem, let's introduce decision variables x_e, one for each edge of the graph. The decision variables can attain values of either 0 or 1. Thus our problem can be written as

$$\text{Minimize } Z = \sum_{e \in M} w_e x_e \tag{A.11}$$

This is our objective function. The requirement that a vertex of $v \in V$ have exactly one incident edge of M is expressed by having the sum of x_e over all edges incident to V equal to one.

$$\sum_{e \in E : v \in V} x_e = 1 \tag{A.12}$$

Thus, we have an Integer Programming Problem (IP):

$$\text{Minimize } Z = \sum_{e \in M} w_e x_e \tag{A.13}$$

$$\text{s.t.} \sum_{e \in E : v \in V} x_e = 1 \tag{A.14}$$

$$\text{and } x_e \in \{0, 1\} \qquad \text{for each edge } e \in E \tag{A.15}$$

If we relax the integer requirement on the decision variables, we have

$$\text{Minimize } Z = \sum_{e \in M} w_e x_e \tag{A.16}$$

$$\text{s.t.} \sum_{e \in E : v \in V} x_e = 1 \tag{A.17}$$

$$\text{and } 0 \leq x_e \leq 1 \qquad \text{for each edge } e \in E \tag{A.18}$$

i) This is called an LP relaxation of the IP.

ii) What good is this?

iii) It leads to a lower bound on the IP optimum solution. Furthermore, sometimes solving the LP relaxation will afford an optimal solution to the previous problem because of total unimodularity.

iv) Theorem: If the LP has a feasible solution, the optimal LP solution will be integral.

A.4.0.5 Algorithm

We will solve this problem with the LP module of the NEOS server.

A.4.0.6 Demonstration

Figure A.7 illustrates the input screen and blocks for the App.

```
Let's say this is the text for your Ampl Assignment model:
param m;
param n;
param c{1..m, 1..n};
var x{1..m,1..n} >= 0;
minimize cost: sum{i in 1..m, j in 1..n} c[i,j]*x[i,j];
subj to supplier{i in 1..m}:
sum{j in 1..n} x[i,j] = 1;
subj to demand{j in 1..n}:
sum{i in 1..m} x[i,j] = 1;
solve;
display x;
And this is the text for your data:

param m:=3;
param n:= 3;
param c: 1 2 3 :=
1 8 6 10
2 9 12 13
3 3 12 11;

The first section of code*:
<document><category>lp</category><solver>CPLEX</solver>
<inputMethod>AMPL</
inputMethod><model><![CDATA[param m;param n;param c{1..m, 1..n};
 var x{1..m,1..n} >=0;
 minimize cost: sum{i in 1..m, j in 1..n} c[i,j]*x[i,j];
 subj to supplier{i in 1..m}:sum{j in 1..n}
x[i,j] = 1;subj to demand{j in 1..n}: sum{i in 1..m} x[i,j] = 1;]]></model><data><![CDATA[
```

A.4.0.7 Evaluation

The App works very well and once the data is input, it solves the problem efficiently.

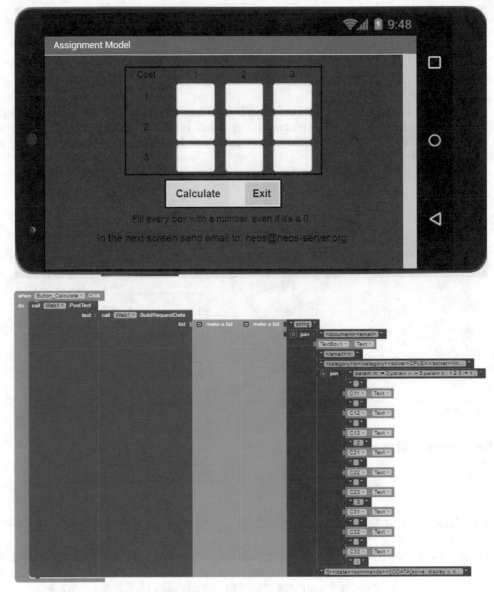

Fig. A.7 Illustration of Assignment Problem Input/Blocks

A.5 Transportation Problem

This problem is closely related to the Assignment problem but its historical development involves many other people unrelated to the Assignment problem. This problem has many applications in production, routing, and manufacturing and service systems.

A.5.0.1 Introduction

The first researcher to study this problem was Tolstoi who in 1930 studied the transportation of salt, cement, and other cargoes along the Soviet rail network. In 1939, Kantorovich developed his approach to the Transportation Problem unaware of the work of Tolstoi. Kantorovich was interested in a number of applications of the problem and his algorithmic approach is very similar to a simplex type approach later developed by Dantzig. He even used dual variables to obtain one alternative approach to the problem which was way ahead of its time.

A.5.0.2 Problem

The problem can be best understood through Figure A.8. We have two supply warehouses and three destinations. The cost of transporting goods is seen in the c_{ij} parameters on the arcs of the graph. We wish to minimize the overall costs for transporting the goods to meet the demands at the destinations.

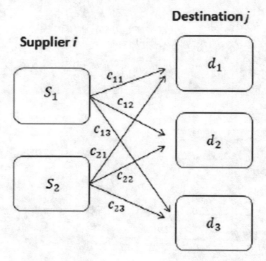

Fig. A.8 Illustration of Transportation Problem

A.5.0.3 Mathematical Model

The input data required for the model include the following: $c_{i}j$ = Unit cost to ship from supplier i to destination j s_i = Capacity of supplier i d_j = Demand of customer j Variables: $x_{ij} \geq 0$: Number of units to ship from supplier i to destination j $i = 1, 2; j = 1, 2, 3$

$$\text{Minimize } Z =: \sum_{i}^{2} \sum_{j}^{3} c_{ij} x_{ij} \tag{A.19}$$

$$Subject\ to: \sum_{1}^{3} x_{ij} \leq s_i\ \forall i \tag{A.20}$$

$$\sum_{1}^{2} x_{ij} \geq d_j\ \forall j \tag{A.21}$$

A.5.0.4 Algorithm

We shall utilize the LP solution algorithm from the NEOS server to demonstrate the solution approach.

A.5.0.5 Demonstration

In the following, we provide the parameters for the App the AMPL mathematical model:

```
Let's say this is the text for your Ampl Transportation model:
param m;
param n;
param s{1..m};
param d{1..n};
param c{1..m, 1..n};
var x{1..m,1..n} >= 0;
minimize shipmentcost: sum{i in 1..m, j in 1..n} c[i,j]*x[i,j];
subj to supplier{i in 1..m}:
sum{j in 1..n} x[i,j] <= s[i];
subj to demand{j in 1..n}:
sum{i in 1..m} x[i,j] >= d[j];
solve;
display x;
And this is the text for your data:
param m:=2;
param n:= 3;
param s:=
1 35
2 50;
param d:=
1 45
2 20
3 30;
param c: 1 2 3 :=
1 8 6 10
2 9 12 13;

The first section of code*:
<document> <category>lp</category> <solver>CPLEX</solver> <inputMethod>AMPL</
inputMethod> <model><![CDATA[param m;param n;param s{1..m};param d{1..n};param
c{1..m, 1..n};var x{1..m,1..n} >= 0;minimize shipmentcost: sum{i in 1..m, j in 1..n}

c[i,j]*x[i,j];subj to supplier{i in 1..m}:sum{j in 1..n} x[i,j] <= s[i];subj to demand{j in
1..n}:sum{i in 1..m} x[i,j] >= d[j];]]></model> <data><![CDATA[
The second section of code**:
]]></data><commands><![CDATA[solve;display x;]]></commands><comments><!
[CDATA[]]></comments></document>
Notice that the first section of code ends right where you will input the data (<data><!
```

A.5.0.6 Evaluation

Figure A.11 illustrates the solution for our example problem. The App does a very good job.

A.6 Shortest Path Problem

This is an extension of the Shortest Path Problem discussed in Chapter 2 where we will solve the problem with Linear Programming (LP) through a network flow formulation (Fig. A.9).

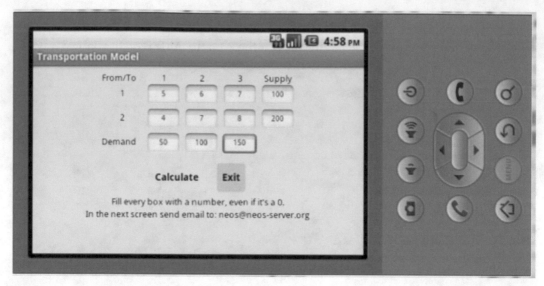

Fig. A.9 Inputs for Transportation Problem

A.6.0.1 Introduction

In many practical situations and with access to LP software, solving the SPT with LP is very reasonable.

A.6.0.2 Problem

The general problem was pretty much discussed in Chapter 2 and will not be repeated here. We want to show how the NEOS server can solve shortest path problems.

A.6.0.3 Mathematical Model

We can formulate the shortest path problem as an LP. This model is based upon a min cost flow model of the shortest path problem. We are going to send one unit of flow from the starting node to the destination node which achieves the minimum cost flow (Figs. A.10 and A.11).

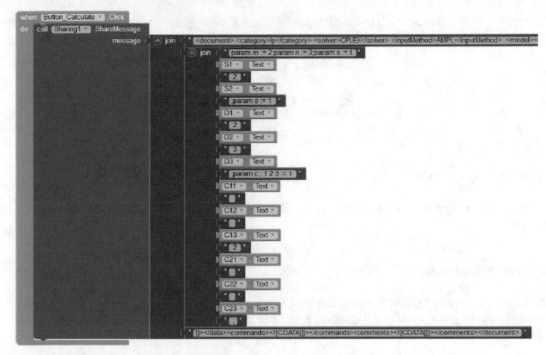

Fig. A.10 Blocks of Transportation Problem

You are using the solver cplexamp.
Checking ampl.mod for cplex_options...
Checking ampl.com for cplex_options...
Executing AMPL.
processing data.
processing commands.
Executing on prod-exec-3.neos-server.org

6 variables, all linear
5 constraints, all linear; 12 nonzeros
        5 inequality constraints
1 linear objective; 6 nonzeros.

CPLEX 12.9.0.0: threads=4
CPLEX 12.9.0.0: optimal solution; objective 2000
3 dual simplex iterations (0 in phase I)
x :=
1 1   0
1 2   100
1 3   0
2 1   50
2 2   0
2 3   150

Fig. A.11 Solution of Transportation Problem

Define

$$x_{ij} := \text{amount of flow in arc(i,j)} \tag{A.22}$$

$$= \begin{cases} 1 & \text{if arc } (i,j) \text{ is on the shortest route} \\ 0, & \text{otherwise} \end{cases} \tag{A.23}$$

$$c_{ij} = \text{length of arc } (i,j) \tag{A.24}$$

Then, the objective function of the LP becomes the following linear objective subject to what are called the conservation of flow equations:

$$\text{Minimize } Z = \sum_{\text{all arcs } \in G} c_{ij} x_{ij} \tag{A.25}$$

$$\sum_{\text{all forward arcs } \in G} x_{ij} - \sum_{\text{all reverse arcs } \in G} x_{ij} = 0 \tag{A.26}$$

$$x_{ij} \geq 0 \tag{A.27}$$

A.6.0.4 Algorithm

The solution algorithm is based on the LP Simplex Method approach after the flow formulation.

A.6.0.5 Demonstration

Below in Figure A.12, we illustrate the input screens for the SPT App. One is a blank screen and the other has data input values for a sample run. In Figure A.13, the blocks for the SPT program are illustrated.

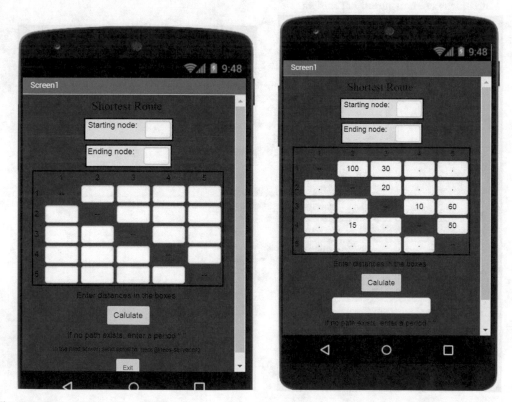

Fig. A.12 Input Screens for SPT Problem

Figure 3.1 is an example we will use to solve the problem starting from node #a to node #g.

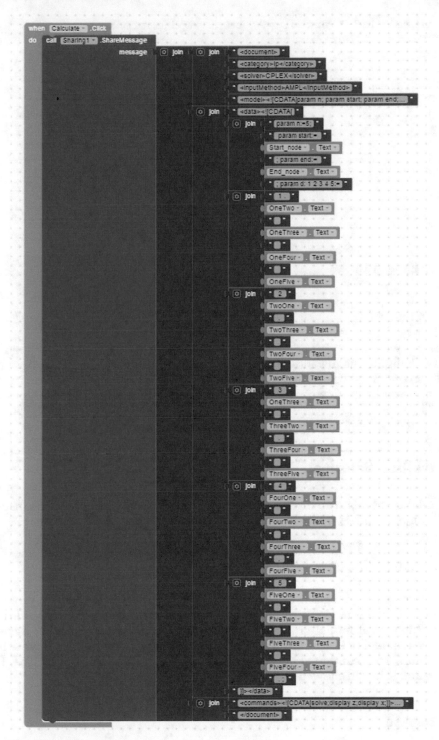

Fig. A.13 Blocks of SPT Problem

```
***BPMPD-AMPL***

BPMPD 2.11: Optimal solution found, objective 140
6 dual value(s) may be wrong; to get correct
```

```
dual values, add " presolv=0" to $bpmpd_options.
0 iterations, 0 corrections

Use :=
a b   1
a c   0
b d   0
b e   1
c d   0
c f   0
d e   0
d f   0
e g   1
f g   0
;

Total_Time = 140
```

A.6.0.6 Evaluation

Works pretty well and it is fairly straightforward and can be expanded to larger graphs simply by changing the distance matrix. Using negative arcs could lead to unbounded LP solutions, so one must be careful with the use of negative arcs. The App is pretty straightforward and works quite well. In the final section, we illustrate the TSP algorithm with NEOS.

A.7 Travelling Salesman Problem (TSP)

The formulation of this App Inventor code is going to be very similar to the Transportation and Assignment models where it has its own formulation and idiosyncracies.

A.7.0.1 Introduction

We have demonstrated a number of examples and approaches to the TSP problem. The TSP history starts out in Germany in 1832 as reported by German traveling salespersons, but no mathematical model or algorithm was developed. In 1930, Menger developed a mathematical model of the problem, and in 1934, H. Whitney is reputed to have solved a 48 city USA tour. Work continued in the 40s at Harvard and Princeton and in 1954, Dantzig, Fulkerson and Johnson showed how LP can be used to solve TSPs.

A.7.0.2 Problem

The TSP problem is $\mathcal{NP}$–Complete for the decision problem which is to find a Hamilton cycle in an arbitrary graph and is $\mathcal{NP}$–Hard for the optimization problem finding the optimum among all Hamilton cycles.

A.7.0.3 Algorithm

Below are the parameters and the mathematical TSP AMPL model.

```
Data
param: city: names :=
1 "1"
2 "2"
3 "3"
4 "4" ;
param DIST: 1 2 3 4 :=
1 500000 132 217 164
2 132 500000 290 201
3 217 290 500000 113
4 164 201 113 500000 ;
AMPL Model
reset;
# Call on CPLEX solver. Note: this line is not needed in App
option solver CPLEX;
set city;
param DIST{city, city} >= 0;
param N := card(city);
param names{city} symbolic;
var U{city} >= 0 integer;
var x{city,city} binary;
minimize z:
sum{i in city, j in city} DIST[i,j]*x[i,j];
subject to
# exactly one outgoing
c1{k in city}: sum{i in city} x[i,k] = 1;
# exactly one incoming
c2{k in city}: sum{j in city} x[k,j] = 1;
# no subtours
c3{k in city, j in city: j > 1 and k > 1}:
U[j] - U[k] + N*x[j,k] <= N-1;
solve;
display z;
display x;
```

A.7.0.4 Demonstration

Figure A.14 illustrates this setup for the TSP.

A.7.0.5 Evaluation

In general, the TSP App does a very good job. Figure A.15 illustrates the programming blocks.

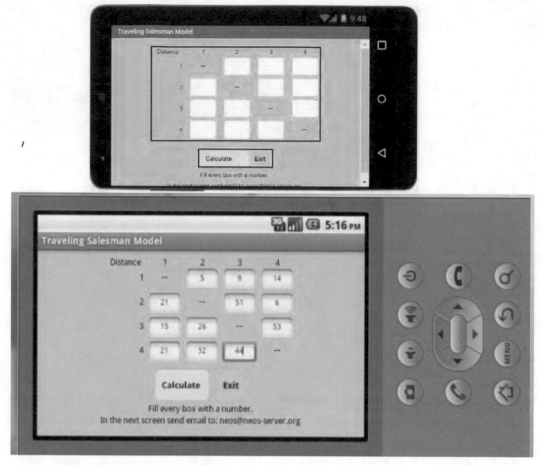

Fig. A.14 Input and Setup of TSP Problem

A.8 Exercises

1. Developing an heuristic solution is often a realistic way to get a close-to-optimal solution in a reasonable amount of computation time. Create an App for the Transportation type problems using Vogel's Approximation Method (VAM) described below for finding a near-optimal solution. It is a penalty type of method. See also the discussion in Taha [14].

   Step 1: For each row (column), determine a penalty measure by subtracting the *smallest* unit cost element in the row (column) from the *next smallest* unit cost element in the same row(column).
   Step 2: Identify the row or column with the largest penalty. Break ties arbitrarily. Allocate as much as possible with the least unit cost in the selected row (column). Adjust the supply and demand, and cross out the satisfied row or column. If a row or column are satisfied simultaneously, only one of the two is crossed out, and the remaining row(column) is assigned zero supply (demand).

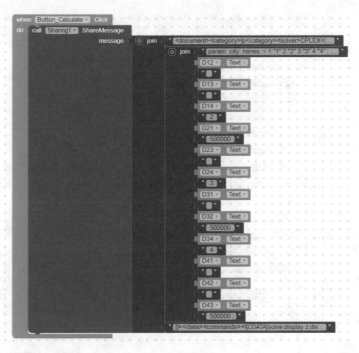

Fig. A.15 Blocks of TSP Problem

Step 3.

(a)   If exactly one row or column with zero supply or demand remains uncrossed, stop.

(b)   If one row(column) with *positive* supply (Demand) remains uncrossed out, determine the basic variables in the row(column) by the least cost method, and stop.

(c)   If all the uncrossed out row and columns have (remaining) zero supply and demand, determine the *zero* basic variables by the least cost method. Stop.

(d)   Otherwise, go to Step 1.

2. **Transportation Problem** Create a complex Transportation Model problem and use the NEOS server for its solution.

3. **Assignment Problem** Create a large scale Assignment problem and run it on the NEOS server.

4. **TSP Problem** Create an optimal tour of your nearest large city or local monument or park system, and model it as a TSP problem. See the Chapter in Taha [14] for some insights.

5. **Max Flow Problem**
   Develop an App for the Maximum Flow Problem. See, for Example, Taha [14].

6. **Min Cost Flow Problem**
   Develop an App for the Min Cost Flow Problem. See for Example Taha [14] (Fig. A.16).

Executing on prod-exec-1.neos-server.org

Presolve eliminates 3 constraints and 4 variables.
Adjusted problem:
16 variables:
     13 binary variables
     3 integer variables
14 constraints, all linear; 44 nonzeros
     8 equality constraints
     6 inequality constraints
1 linear objective; 13 nonzeros.

CPLEX 12.9.0.0: threads=4
CPLEX 12.9.0.0: optimal integer solution; objective 62
4 MIP simplex iterations
0 branch-and-bound nodes
z = 62

x :=
1 1 0
1 2 0
1 3 1
1 4 0
2 1 0
2 2 0
2 3 0
2 4 1
3 1 0
3 2 1
3 3 0
3 4 0
4 1 1
4 2 0
4 3 0
4 4 0

Fig. A.16 Solution of TSP Problem

Glossary

AI2   App Inventor Programming Language, Version II

CDATA   Software in email to NEOS for data inputs from AMPL

CPLEX   Software for large scale mathematical programs

G(V,E)   Graph with finite set of Vertices V and Edges E

M/G/c/c   Markovian arrivals, general service c-servers, c-waiting room, no queueing allowed

DTM   Deterministic Turing Machine, after Alan Turing, model for today's computers.

NDTM   Non-deterministic Turing Machine, fictitious machine, capable of traversing all solutions paths in polynomial time.

$\mathcal{N}\mathcal{P}$ Class of problems solvable on a Non-deterministic Turing Machine in polynomial time.

$\mathcal{N}\mathcal{P}$– **Complete**   Class of problems reducible to the three satisfiability problem but not solvable in polynomial time on a DTM, (*i.e.* Hamilton Cycle)

$\mathcal{N}\mathcal{P}$– **Hard**   Class of problems in $\mathcal{N}\mathcal{P}$ which are also optimization problems

$\mathcal{P}$ Class of problems solvable in Polynomial time on a DTM

© Springer Nature Switzerland AG 2021
J. MacGregor Smith, *Combinatorial, Linear, Integer and Nonlinear Optimization Apps*,
Springer Optimization and Its Applications 175,
https://doi.org/10.1007/978-3-030-75801-1_A

References

[1] Chong, E.K.P., Zak, S.H.: Introduction to Optimization, 4th ed. Wiley (2013)

[2] Pierre, D.: Optimization Theory with Applications. Dover Books (1985)

[3] Churchman, C.W., Ackoff, R.L.: Introduction to Operations Research. Wiley, New York (1957)

[4] Cohon, J.: Multiobjective Programming and Planning. Academic, New York (1978)

[5] Dantzig, G., Fulkerson, R., Johnson, S.M.: Solution of a large scale traveling salesman problems. Oper. Res. **2**(4), 393–410 (1954)

[6] Deleforge, A.: The maths of group testing: mixing samples to speed up COVID-19 testing (2020). https://members.loria.fr/ADeleforge

[7] Fourer, R., Gay, D.M., Kernighan, R.: AMPL: a modeling language for mathematical programming

[8] Johnson, S.M.: Optimal two- and three-stage production schedules with set-up time included. Nav. Res. Logist. Q. **1**, 61–68 (1954)

[9] Kamriani, F., Roy, K.: App inventor 2 essentials

[10] Lawler, E.: Combinatorial Optimization: Networks and Matroids. Holt, Rinehart, and Winston (1976)

[11] Rittel, H., Webber, M.: Dilemmas in a general theory of planning. Policy Sci. **4**, 155–167 (1973)

[12] Russel, D.: Optimization Theory. W.A. Benjamin, Inc., New York (1970)

[13] Schrijver, A.: Combinatorial Optimization. Springer, Berlin (2003)

[14] Taha, H.: Operations Research: an Introduction, 10th edn. Pearson, New Jersey (2018)

[15] Winston, W.: Introduction to Probability Models, 4th edn. Thomson Brooks/Cole, Belmont (2004)

[16] Wolber, D.: www.appinventor.org/book2

[17] Woolsey, R.E.D., Swanson, H.S.: Operations Research for Immediate Application: a Quick and Dirty Manual. Harper and Row, New York (1975)

© Springer Nature Switzerland AG 2021
J. MacGregor Smith, *Combinatorial, Linear, Integer and Nonlinear Optimization Apps*,
Springer Optimization and Its Applications 175,
https://doi.org/10.1007/978-3-030-75801-1_A

Index

© Springer Nature Switzerland AG 2021
J. MacGregor Smith, *Combinatorial, Linear, Integer and Nonlinear Optimization Apps*,
Springer Optimization and Its Applications 175,
https://doi.org/10.1007/978-3-030-75801-1

Printed in the United States
by Baker & Taylor Publisher Services